Wetlands for the Treatment of Agricultural Drainage Water

Wetlands for the Treatment of Agricultural Drainage Water

Special Issue Editor

Guangzhi Sun

MDPI • Basel • Beijing • Wuhan • Barcelona • Belgrade

MDPI

Special Issue Editor
Guangzhi Sun
Edith Cowan University
Australia

Editorial Office
MDPI
St. Alban-Anlage 66
Basel, Switzerland

This is a reprint of articles from the Special Issue published online in the open access journal *Water* (ISSN 2073-4441) in 2018 (available at: http://www.mdpi.com/journal/water/special_issues/ Wetlands_Agricultural_Drainage_Water)

For citation purposes, cite each article independently as indicated on the article page online and as indicated below:

LastName, A.A.; LastName, B.B.; LastName, C.C. Article Title. *Journal Name* **Year**, *Article Number*, Page Range.

ISBN 978-3-03897-208-2 (Pbk)
ISBN 978-3-03897-209-9 (PDF)

Contents

About the Special Issue Editor

Guangzhi Sun, Associate Professor, is an academic in the School of Engineering at Edith Cowan University, Australia. He received his PhD in Chemical Engineering from The University of Birmingham in the UK, and MEng and BEng from Tianjin University in China. He has over twenty years of research experience in the field of constructed wetland for water pollution control, having published over sixty journal papers that cover the removal of organics, nutrients, metals and metalloids in wetland systems, process modelling, and wetland hydrology.

Preface to "Wetlands for the Treatment of Agricultural Drainage Water"

Natural wetlands are known as the 'kidneys' of the earth and are important aquatic systems where water self-cleaning processes take place. Centuries ago, in ancient Egyptian and Chinese civilisations, people already recognised that dirty waters can be somewhat 'cleaned' by channelling and draining them to marshes and swamps. Constructed wetlands are a relatively new concept, developed in northern Europe in the mid-twentieth century. They are manmade systems, hence 'artificial kidneys', purposely built in the vicinity of a source (or sources) of pollutants, to intercept and remove the pollutants. Constructed wetlands usually serve as a part of wastewater and stormwater management infrastructures. Today, natural and constructed wetlands all play a critical role in the health of our ecosystems and environment protection.

Agricultural drainage typically includes irrigation waters from paddy fields and runoffs from wineries and animal farms; many have elevated concentrations of organics and nutrients that present a pollution threat to the water environment. It is unacceptable in developed communities to simply 'get rid of' these waters (i.e., untreated discharge), even though few options are available to effectively manage them. Due to significant pollutant input and seasonal variations, agricultural drainage system must be able to maintain its own structure, immobilise the pollutants, sustain sufficient biotic and/or physicochemical processes to permanently remove the pollutants and prevent secondary pollution, and be economically viable. To date, this remains a technical challenge that the research communities have more or less failed to tackle.

For effective management of agricultural drainage, a number of rural environmental issues need to be addressed simultaneously, such as sustainable use of land and water resources, hazardous and non-hazardous waste management, and wastewater treatment. Controlled discharge and treatment in wetlands can potentially become a major part of an integrated solution to the problem of agricultural drainage. To this end, some key questions to be answered include: What is the maximum pollutant load (especially nutrients) that a certain type of wetland can cope with? What is the fate of the pollutants retained in the wetlands? How to predict long-term performances from short- or medium-term data? Can some non-hazardous agricultural waste be used as local materials to establish constructed wetlands? and what are the suitable engineering interventions to enhance the functionality of the wetlands? This special edition includes some up-to-date studies on these topics, to help develop wetland technology closer towards solving the agricultural drainage problem.

The publication of this special edition could not have been achieved without the generous support of many people. In particular, I would like to express my gratitude to the following people involved.

Authors of all the papers, for their contributions and meticulous revisions undertaken during the peer review process;

All the reviewers, for their selfless assistance to safeguard the quality of these papers;

Rachel and Lynette in the Water Editorial Office, for their endless patience and support;

MDPI, for giving me the opportunity.

Guangzhi Sun
Special Issue Editor

water

MDPI

Article

Effectiveness of a Natural Headwater Wetland for Reducing Agricultural Nitrogen Loads

Evelyn Uuemaa [1,2,*], Chris C. Palliser [2], Andrew O. Hughes [2] and Chris C. Tanner [2]

[1] Department of Geography, University of Tartu, 51014 Tartu, Estonia
[2] National Institute of Water and Atmospheric Research Limited, P.O. Box 11 115,
Hamilton 3251, New Zealand; chris.palliser@niwa.co.nz (C.C.P.); andrew.hughes@niwa.co.nz (A.O.H.);
chris.tanner@niwa.co.nz (C.C.T.)
* Correspondence: evelyn.uuemaa@ut.ee; Tel.: +372-737-5827

Received: 10 February 2018; Accepted: 2 March 2018; Published: 8 March 2018

Abstract: Natural wetlands can play a key role in controlling non-point source pollution, but quantifying their capacity to reduce contaminant loads is often challenging due to diffuse and variable inflows. The nitrogen removal performance of a small natural headwater wetland in a pastoral agricultural catchment in Waikato, New Zealand was assessed over a two-year period (2011–2013) Flow and water quality samples were collected at the wetland upper and lower locations, and piezometers sampled inside and outside the wetland. A simple dynamic model operating on an hourly time step was used to assess wetland removal performance for key N species. Hourly measurements of inflow, outflow, rainfall and Penman-Monteith evapotranspiration estimates were used to calculate dynamic water balance for the wetland. A dynamic N mass balance was calculated for each N component by coupling influent concentrations to the dynamic water balance and applying a first order areal removal coefficient (k_{20}) adjusted to the ambient temperature. Flow and water quality monitoring showed that wetland was mainly groundwater fed. The concentrations of oxidised nitrogen (NO_x-N, Total Organic Nitrogen (TON) and Total-N (TN) were lower at the outlet of the wetland regardless of flow conditions or seasonality, even during winter storms. The model estimation showed that the wetland could reduce net NO_x-N, NH_4-N, TON and TN loads by 76%, 73%, 26% and 57%, respectively.

Keywords: wetland attenuation; nitrogen; nutrient removal; denitrification; modelling; agricultural pollution

1. Introduction

Nitrogen loads from diffuse water pollution (DWP) are a major water quality problem in many countries [1]. Compared to point source pollution, DWP is more complex and difficult to control due to its numerous and dispersed sources, and the difficulties in tracing its pathways [2]. Wetlands have been demonstrated to be an effective means to attenuate nitrogen derived from DWP [3,4] by plant and periphyton uptake and microbial denitrification [5,6]. In some wetlands, studies have shown that denitrification is the dominant nitrate removal mechanism [4,7,8]. However, microbial activity, which controls denitrification rates, can be reduced at low temperatures [9], pH, and carbon availability [6,10]. In line with these seasonal differences in wetlands, nitrogen removal performance is widely reported with reduced rates measured at colder temperatures [4]. Bernal [11] examined N content, N accumulation rates and soil C:N ratios over time in two riverine wetlands in the U.S. Midwest and found that, besides denitrification, organic accumulation was also important in N removal. Zaman [12] found that denitrification only accounted for 6–7% of observed NO_3 removal in a short-term injection-resampling study, and suggested that plant uptake was the principal removal mechanism.

In New Zealand, about half of the land area is used in some form of pastoral production, ranging from intensively farmed lowland (often dairy) to extensively farmed hill country (usually sheep/beef/deer) [13]. Dairy cow numbers have risen from approximately 3 million in 1980 to approximately 6.5 million in 2015 [14] along with increased application of fertilisers and use of supplementary feed. For example, N fertiliser use increased from 50 Gg in 1989 to 329 Gg in 2010 [15]. At the same time, extensive drainage across many parts of New Zealand have converted large areas of natural wetlands into agricultural land [3] and small remnant wetland areas continue to be drained as farming practices intensify [16]. In New Zealand, it has been estimated that over 90% of the former wetland area has been lost within a century and a half [17] and the trend is continuing particularly for small wetlands in agricultural landscapes. Ongoing intensification has raised concerns about environmental sustainability and contamination of groundwater and surface water with nutrients particularly under intensive dairying [18,19]. Therefore, there is increasing need for effective management tools to reduce the nutrient losses to water bodies [13].

Natural headwater wetlands are a relatively common feature in the hilly parts of New Zealand. These wetlands often occur within the headwater areas of catchments and along the sides of streams [20]. Although they are individually small, they may represent a significant proportion of headwater catchments. The potential of these wetlands to attenuate upslope derived pollutants is well recognised [21]. However, they can also be a potential source of agricultural pollutants because of their direct connection to the stream network, and farmers see them as a suitable drinking water source for livestock [20]. Pastoral wetlands are often small (less than 5000 m^2) and therefore they are rarely identified in wetland inventories or managed, and the ecosystem services they provide are poorly understood. Compared to constructed wetlands, the contaminant removal processes in natural wetlands are generally more complicated to study because they have diffuse, spatially and temporally variable groundwater inflows and outflows that are difficult to access and measure. There have been few attempts to quantify the effectiveness of natural seepage wetlands [22], whereas nutrient removal of constructed wetlands with discrete inflows and outflows is comparatively well studied [3,23–26].

Models are being increasingly used to help quantify and predict the efficacy of wetlands for removing excess nutrients. The biological, chemical and physical removal processes in wetlands vary in space and time [11]. Modelling enables improved understanding of the complex interplay of hydrology and biogeochemical processes taking place in wetlands and their variability in space and time at relatively low cost.

This study aims to estimate the performance of a natural headwater wetland in removing nitrogen loads originated from upland grazed dairy pasture. It was hypothesised that wetland N removal efficiency would vary depending on inflow loads and seasonal temperatures, and that the coincidence of high flows and low temperatures in winter would reduce removal efficiencies. The performance of the wetland was estimated by assessing nitrogen in-loads and out-loads by flow and water quality monitoring and modelling.

2. Materials and Methods

2.1. Study Site

The study wetland is located on a dairy farm near Kiwitahi in the headwaters of the Toenepi catchment (15.8 km^2) in the eastern Waikato region, New Zealand (Figure 1). The Toenepi catchment is intensively farmed and 75% of the catchment area is under dairy production with a stocking rate of ~3 cows ha^{-1} [27]. The mean annual rainfall of the area is 1377 mm. The upper Toenepi catchment is hilly with ~80% of the area classified as either rolling or steep (>10% gradient) [28]. The wetland catchment is dominated by Morrinsville clay soil (NZ Soil Classification: Orthic Granular).

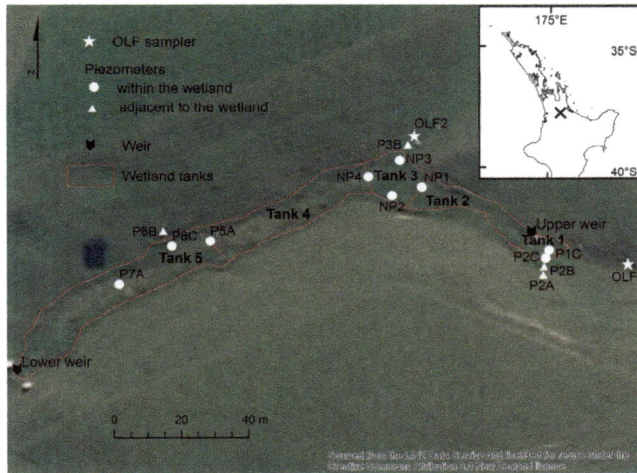

Figure 1. Wetland study site location in the North Island of New Zealand, showing sampling weirs, piezometers, overland flow samplers and conceptual tanks used for modelling of wetland hydrology and N attenuation.

Rotational grazing of a single herd of ~220 (Holstein Friesian) cattle is practiced on the dairy farm, which is divided into 33 individual paddocks or fields (fenced pasture area for grazing) of between 1.0 and 3.1 ha. The study wetland is located within a small (~1.9 ha) fenced paddock, which is grazed for ~1 day every 40 days during winter and summer and ~1 day every 20 days during spring and autumn. As most of the time the wetland does not provide ready access to surface water, drinking water is available to the herd from a water trough (groundwater bore source) within the paddock. The paddock containing the study wetland (with exception of the wetland itself) is steep (mostly exceeding 20° slope). The extent and impacts of grazing events on water quality at the outlet of the wetland over the period of the present study have been described by Hughes [29].

The wetland has an area of ~0.15 ha (2.8% of surface catchment) and an average slope of 3.5°. The permanently saturated wetland soil is composed largely of a mix of organic material and the clay-based soils eroded from the surrounding hillslopes. Sediment probe measurements indicated that within 1 m of the edge the saturated layer was generally between 0.5 and 1 m thick. Depths increased with distance from the edge, and were generally between 1 and 2 m in the centre of the wetland. The wetland vegetation is dominated by glaucous sweet grass (*Glyceria declinata*), a perennial aquatic grass widely naturalised in New Zealand.

2.2. Site Monitoring

The study site was monitored for a two-year period between October 2011 and September 2013. Flow was measured hourly at two 45° v-notch weirs with stage height measured by a Unidata Hydrologger water level recorder (1 mm resolution; Unidata Pty, O'Connor, WA, Australia) and converted to flow using a theoretical rating equation. The Upper Weir (UW) was located at the head of wetland, downstream from where there was significant ground water seepage (Figure 1). The catchment area above the UW is ~2.9 ha. The Lower Weir (LW) was located within a constricted part of the lower wetland with a total catchment area of ~5.2 ha (Figure 1). During the period between November 2012 and May 2013, the study site experienced exceptionally dry conditions [30]. Consequently, no flow was recorded at the UW from early January 2013 through to May 2013. Despite this, the wetland remained wet and boggy with flow at the LW.

An ISCO 3700 automatic water sampler (Teledyne Isco, Lincoln, NE, USA) was programmed to collect water samples behind each weir using a stage-based trigger. In addition, low flow grab samples were collected during site visits approximately every six weeks. Once collected, samples were immediately placed in an insulated storage bin containing an ice slurry. Samples were delivered to the NIWA—Hamilton Water Quality Laboratory on the day of collection for grab samples and within 24 h for samples collected from the automatic sampler.

Piezometers were installed at 13 locations, both within (nine piezometers) and adjacent to the wetland (four piezometers) (Figure 1). The within-wetland piezometers were installed to below the depth of the deposited wetland material (i.e., into the weathered underlying bedrock material) and were slotted to collect water from depths of 1–2.5 m. During site visits for grab samples, piezometer water level was recorded and water samples were collected. Two overland flow (OLF) samplers were also monitored at the site, one in the gully immediately upstream of UW (OLF1) and one at the base of a swale on the northern margin of the wetland (OLF2).

Rainfall was also measured at the lower weir at 10 min intervals by an Ota tipping bucket rain gauge (Ota Keiki Seisakusho, Tokyo, Japan). Potential evapotranspiration was calculated using the FAO Penman–Monteith equation [31].

2.3. Laboratory Analysis

Collected water samples were analysed for, oxidised N (nitrite- and nitrate-N, here-after referred to as NO_x-N), ammonium-N (NH_4-N), Total Nitrogen (TN). Total Organic Nitrogen (TON) was calculated by subtraction (TN minus (NO_x-N plus NH_4-N)). A Lachat flow injection analyser (Hach, Loveland, CO, USA) was used for NO_x-N, NH_4-N (detection limit 1 mg/m^3), and for TN (detection limit 10 mg/m^3). All water samples were filtered after subsampling for TN (as well as total suspended solids and total phosphorus not reported here) with a Millipore® syringe and filter holder containing a GF/C glass fibre pre-filter (47 mm diam., 1.2 μm pore size), and a Sartorius® cellulose acetate membrane filter (47 mm diam., 0.45 μm pore size).

To determine whether pollutant concentrations varied by season, the Kruskal–Wallis statistical test was used to test for differences in median seasonal pollutant concentrations. The Kruskal–Wallis test can be considered to be a non-parametric version of the ANOVA test. The Mann–Whitney U test was used for comparing upper and lower weir median pollutant concentrations. A significance level of $p < 0.05$ was adopted in all tests. The statistical software package Statistica 12 (StatSoft; Tulsa, OK, USA) [32] was used to perform these tests.

2.4. Modelling Nitrogen Removal

We used N measurements and flow measurements as inputs to a simple dynamic model [32,33] to explore potential wetland N removal performance during 2012 when continuous flow data were available. The model was set up in Microsoft Excel (365 Pro Plus ver. 1708, Redmond, WA, USA) (Figure 2) and uses Euler integration (Equation (1)). To simulate the internal hydraulic dynamics of the wetland, we used a five tanks-in-series approach (Figures 1 and 2) with areas from 56 m^2 to 637 m^2 (Table 1). Tank 1 was situated above UW and Tank 5 terminated at LW. Each tank had at least one piezometer located near its outlet. Hourly measurements of flow at UW and LW, rainfall, and Penman evapotranspiration estimates were used to calculate a dynamic water balance for the wetland.

Table 1. Summary of wetland tank characteristics.

	Wetland Area (m²)	Apparent Catchment Area * (m²)	Over 15° Slope Areas Draining to the Wetland (m²)
Tank 1	56	23,710	3613
Tank 2	191	4408	803
Tank 3	173	14,397	1686
Tank 4	409	4697	1021
Tank 5	637	4341	1096
Total	1467	51,553	8218

Note: * Excluding areas draining to upstream tanks.

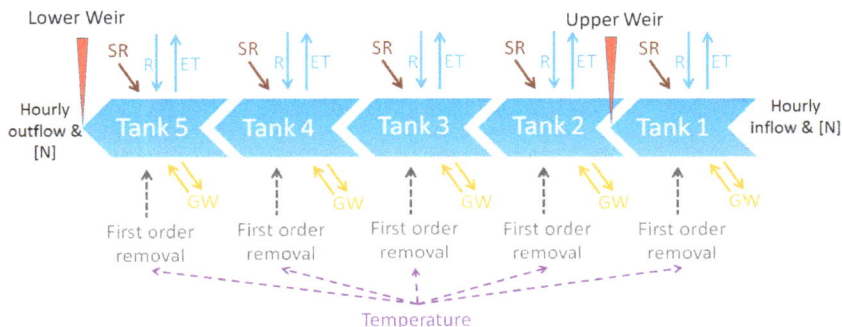

Figure 2. Conceptual diagram of the wetland model. Rainfall (R), evapotranspiration (ET), groundwater (GW) seepage in or out based on each tank's area, surface runoff (SR), and first order N removal adjusted by ambient temperature are calculated for each tank on an hourly time step.

The following differential equation (Kadlec, 2012) was used for the dynamic water balance:

$$\frac{d(V_{i+1})}{dt} = Q_i - Q_{i+1} + A_{i+1}(P_{i+1} - ET_{i+1}) + SR_{i+1} + GW_{i+1}, \tag{1}$$

where t is time (h), $i+1$ refers to Tank$_{i+1}$, i refers to Tank$_i$, V is the volume of water in the tank (m³), Q is the flow (m³/h), A is the surface area for the tank (m²), P is the precipitation (m/h), ET is the evapotranspiration (m/h), SR is the surface runoff inflow (m³/h), and GW is the net groundwater inflow (m³/h). When weir is not present, then Q_i is calculated from water balance based on inputs/outputs from neighboring tanks, precipitation, surface runoff and evapotranspiration calculated based on each tank's area.

Storm events above a certain threshold (rainfall intensity ≥ 1 mm/h, flow at weirs increasing and storm length ≥ 3 h) were assumed to always result in surface runoff. Altogether, 37 such storms were defined over the period of this study (12 months, January 2012 to December 2012), the shortest being 3 h and the longest 27 h. SR into each tank was calculated as P-ET during the storm event and multiplied by the catchment area of the tank. The subsurface infiltration rate of groundwater into or out of the wetland was derived from the water balance calculated at the outlet of the wetland, assuming that the wetland is a closed system. Changes in wetland depth and hence storage volume were simulated for a 45° V-notch weir set 35 cm above the base of the wetland using a modification of the Francis weir formula [34].

A dynamic N mass balance was calculated by coupling influent concentrations to the dynamic water balance and applying a first order areal removal rate coefficient (k) (Equation (2)) [35].

The mass balance for a tank is given by the differential equation:

$$\frac{d(V_{i+1}C_{i+1})}{dt} = Q_i C_i - Q_{i+1}C_{i+1} + A_{i+1}(C_P P_{i+1} - C_{ET}ET_{i+1} - (C_P - C_{ET} - C^*)k), +SR_{i+1}C_{SR} \\ +GW_{i+1}C_{GW}, \quad (2)$$

where C is concentration (mg N/m^3) and k is the removal rate coefficient (m/h).

The removal rate coefficient (k) is well known to be temperature sensitive [35]. The modified Arrhenius equation was used for adjusting k to the ambient temperature with temperature factors derived from field-scale wetland studies (Table 2) [35,36]. The annual water temperature regime was simulated using a sinusoidal function calibrated to air temperatures measured at the site, as described by Kadlec [33]. We adjusted within the range of k_{20} coefficients documented for surface flow wetlands (Table 2). For modelling net TON and TN removal, an irreducible background concentration C^* was used (Table 2; Equation (2)).

Table 2. Ranges for removal rate coefficients (k), temperature factors (θ) and background concentrations used for modelling NH$_4$-N, NO$_x$-N, TON and TN removal.

	k_{20}	C^* (mg/m^3)	θ	Reference
NH$_4$-N	5–86	0	1.049	Kadlec and Wallace [35]
NO$_x$-N	5–168	0	1.106	Kadlec [4]
TON	5–62	20	1.017	Wilcock [36]
TN	4–40	200	1.056	Kadlec and Wallace [35]

Surface runoff was assigned higher NH$_4$-N, NO$_x$-N, TON and TN concentrations than UW, based on limited measurements of surface runoff around the wetland (Table 3). The median value recorded for each tank was used as constant input value for groundwater NH$_4$-N, NO$_x$-N, TON and TN (Table 3). Atmospheric deposition of dissolved inorganic forms of N in New Zealand is considerably lower than commonly found in continental Northern Hemisphere regions [33,37], ranging from 1–5 g N ha^{-1} year^{-1} [38]. Therefore, a very low constant input value (10 mg/m^3) for all forms of nitrogen from rainfall was used. To improve the stability of the model, we decreased the time-step during the storms to 10 min. Wetland performance was calculated as:

$$\text{Removal efficiency (\%)} = (\text{Inload} - \text{Outload})/\text{Inload} \times 100\%, \quad (3)$$

and areal mass load removal rate was calculated:

$$\text{Areal mass load removal rate (mg/m}^2/\text{day)} = (\text{Inload} - \text{Outload})/\text{wetland area}/\text{inflow days.} \quad (4)$$

Table 3. Median and min-max concentrations of forms of nitrogen by sampling site.

Site	n	NH$_4$-N (mg/m^3)	NO$_x$-N (mg/m^3)	TON (mg/m^3)	TN (mg/m^3)
Piezometers adjacent to the wetland	15	99 (27–389)	55 (1–885)	441 (148–3266)	688 (235–3850)
Piezometers within the wetland	44	17 (2–2440)	956 (1–3930)	461 (139–5927)	2495 (533–10700)
Overland-flow (OLF)	6	228 (39–1490)	3165 (367–4860)	3929 (2134–8352)	7320 (4830–13,500)
Upper weir					
Baseflow	11	46 (5–733)	186 (4–923)	628 (266–1591)	880 (640–2510)
Summer/Autumn	9	105 (55–487)	934 (346–1330)	2640 (2005–4263)	3851 (2630–6080)
Winter/Spring	36	65 (40–110)	715 (447–1680)	2383 (1247–3595)	3245 (2490–4360)
Lower weir					
Baseflow	15	14 (5–177)	4 (1–52)	322 (117–1594)	381 (129–1600)
Summer/Autumn	33	43 (12–537)	28 (1–376)	586 (387–4601)	864 (550–4830)
Winter/Spring	47	22 (8–67)	280 (1–1450)	1009 (167–1768)	1490 (321–2130)

3. Results

3.1. Hydrology of the Wetland

Annual precipitation of 1056 mm was recorded for the wetland catchment in 2012. About 40% of the rainfall occurred during winter, 30% in spring, 20% in summer and 10% in autumn (Figure 3). During 2012, flow occurred on all days at the LW (lower weir) but only on 122 days at the UW (upper weir). During that 122-day period, there was 1836 m^3 inflow to the wetland as measured at the UW and 12,439.6 m^3 outflow from the wetland as measured at the LW. Flow measured at the upper weir accounted for less than 15% of the flow exiting the lower weir. This is despite the area above the upper weir contributing over half the catchment area and including a highly convergent gully immediately upstream from the wetland. This suggests that a large proportion (85%) of the flow exiting the wetland is derived from subsurface inflows entering between the upper and lower weirs. Subsurface flow appeared to dominate regardless of conditions and time of year. However, it is difficult to accurately estimate the proportion of the seepage because it was not possible to accurately measure all surface runoff inputs during storm events. The estimated inputs to the wetland during 2012 were: 94.7% via groundwater (11,781.6 m^3), 4.1% via surface runoff (511.4 m^3), 1.2% via rain-ET (146.5 m^3).

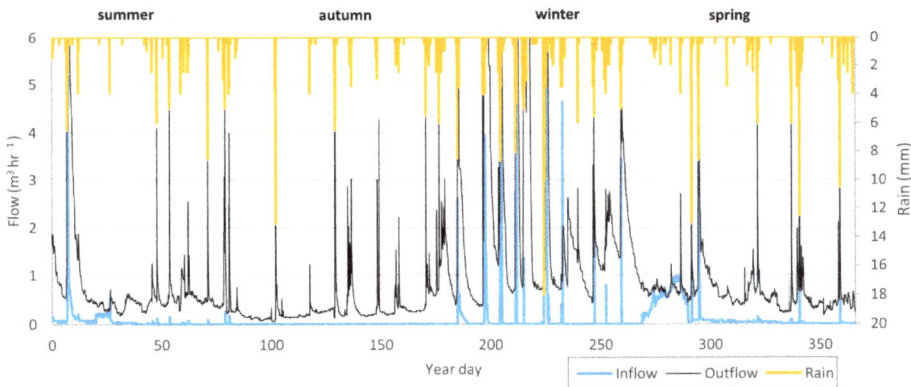

Figure 3. Measured rainfall, upper weir inflow and lower weir outflow of the wetland (with Southern Hemisphere seasons indicated). Year day = 0 corresponds to 1 January 2012. Inflow extends to a maximum of 74.5 m^3/h and outflow to 197.6 m^3/h, which occurred during an extreme storm in 23 July (year day = 205).

3.2. Water Quality of the Wetland

Altogether, 56 water quality samples from the upper weir and 95 samples from the lower weir were analysed (Table 3, Figure 4). Baseflow samples were collected throughout the study period and flow events were sampled over a range of both summer and winter conditions. Water quality data for storms has been grouped together into summer/autumn (December–May) and winter/spring (June–November).

When comparing upper and lower weir, all the measured variables showed lower (NO_x-N and TN statistically significantly, $p < 0.05$) concentrations at the lower weir, regardless of flow conditions or seasonality (Figure 4, Table 2). During the winter/spring storms, all measured variables had significantly higher values ($p < 0.05$) at the upper weir (Tables 3 and 4).

In terms of flow conditions, the concentration of all variables was highest during storm events (Figure 4). At the upper weir, all stormflow variables were highest during summer, which could be due to more frequent stock access to the wetland paddock during this period [29]. However, at the lower weir, all variables, except NH_4-N, were highest for winter storms.

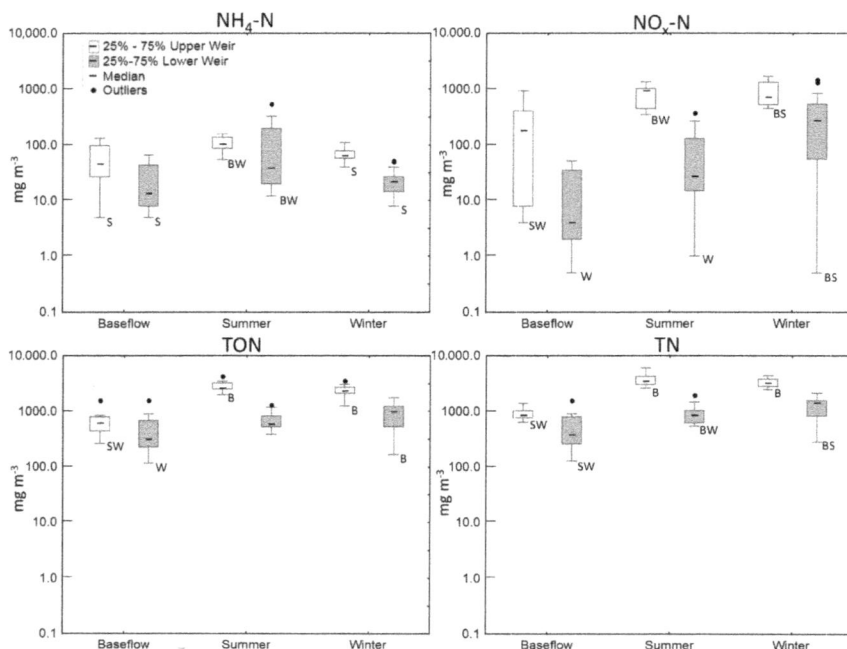

Figure 4. Box plots of different forms of N during baseflow, summer storm flow and winter storm flow at the upper (unshaded boxes) and lower (shaded boxes) wetland weirs. The 'summer' period is defined as the period between the months of December and May while 'winter' is defined as June through to November. Based on Kruskal–Wallis test and post-hoc multiple comparisons: B—statistically significant difference from baseflow, S—statistically significant difference from summer storms, W—statistically significant difference from winter storms ($p < 0.05$).

Table 4. Comparison of upper and lower weir nitrogen variables. Mann–Whitney U Tests statistics shown in bold are significant at $p < 0.05$.

Variable	Baseflow		Summer Storms		Winter Storms	
	Z	*p*-Value	Z	*p*-Value	Z	*p*-Value
NH_4-N	1.54	0.12	1.27	0.20	**7.52**	**0.00**
NO_x-N	**3.24**	**0.00**	**4.52**	**0.00**	**5.72**	**0.00**
TDN	**2.33**	**0.02**	**4.26**	**0.00**	**7.02**	**0.00**
TON	1.57	0.12	**4.28**	**0.00**	**7.32**	**0.00**
TN	**2.49**	**0.01**	**4.31**	**0.00**	**7.77**	**0.00**

Fifty-nine groundwater samples were analysed. Fifteen of those were collected adjacent to the wetland area and 44 within the wetland. Only six discreet overland flow samples were collected from two sites, but, during storm-flows, OLF represented most of the flow at the upper weir. This confirmed that OLF entering the wetland had the most elevated levels of all measured pollutants (Figure 5). Groundwater around the wetland always had relatively low N concentrations. In case of NO_x-N, it was significantly lower ($p < 0.05$) than overland flow and piezometers sampled within the wetland. Despite the similarity of the nitrate concentrations from around the wetland, the NO_x-N concentrations from the sample sites within the wetland showed remarkable spatial variation. Above the UW, the measured NO_x-N concentrations were consistently higher than 3000 mg/m^3. Higher NO_x-N concentrations were also detected in piezometer NP3, which was sited where another gully entered the wetland from north downstream of the UW. NH_4-N and TON concentrations were significantly lower in the wetland groundwater than in OLF.

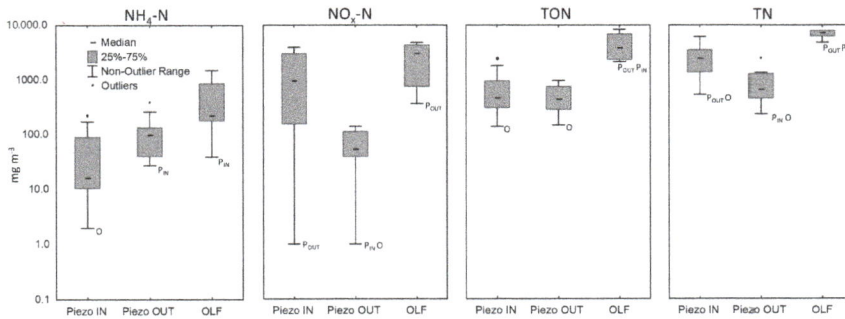

Figure 5. Box plots of different forms of N from piezometers within the wetland (Piezo IN), piezometers adjacent to the wetland (Piezo OUT) and overland flow samplers (OLF). Based on Kruskal–Wallis test and post hoc multiple comparisons: P_{IN}—statistically significant difference from piezometers within wetland, P_{OUT}—statistically significant difference from piezometers adjacent to the wetland, O—statistically significant difference from overland flow ($p < 0.05$).

3.3. Modelling Wetland N Species Removal Efficiency

NO_x-N was the dominant form of N (60.7% of in load) entering the wetland (Figure 6) mainly via groundwater seepage (90.7%). TON also comprised a significant proportion of the N loading (29.9% of in loads) with NH_4-N only 9.4% of the load. TON and NH_4-N were also mainly entering via groundwater (76.8% and 95.1%, respectively). However, a significant portion of TON was also entering the wetland via surface runoff (23%). TON was the dominant form leaving the wetland (56.1% of out load) followed by NO_x-N (37.5% of out load) and NH_4-N (6.4%).

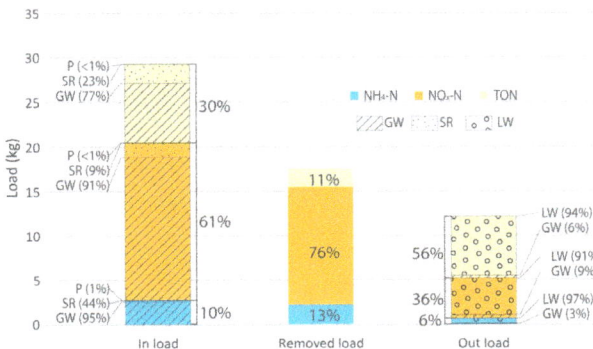

Figure 6. Proportions of total nitrogen components entering and leaving the wetland, and removed. P = precipitation; SR = Surface runoff; GW = Groundwater; LW = Lower weir.

Best fit k_{20} values based on least squares varied for nitrogen forms. NO_x-N and TN showed best fit at the upper boundary of commonly recorded k_{20} rates (168 and 40 m year^{-1}, respectively; Table 2) and TON at lowest levels (5 m year^{-1}). Calculated wetland removal efficiency (Equation (3)) varied considerably for different nitrogen forms (Table 5). Modelling results show substantial reductions in NH_4-N and NO_x-N loads after passage through the wetland, except during the winter period when load reductions were more muted (Figure 7a,b; Table 5). Although the overall NH_4-N and NO_x-N mass removal efficiency for the wetland (Equation (3)) was 72.9% and 75.5%, respectively, the removal efficiency during the winter storms was only 43.5% for NH_4-N and 67.3% for NO_x-N. This is likely to result from the coincidence of high flows (resulting in reduced hydraulic residence times; HRT) and low water temperatures during winter (Figure 7d). However, despite reduced removal efficiency NO_x-N and NH_4-N mass removal was actually greater over the winter period because of the higher and more consistent loads received (Table 5). The modelling showed only 26% of net TON reduction in the wetland, with highest rates during summer/autumn storm events. Annual reduction of TN was 57.2%, varying from 39.8% during baseflow to 75% during summer/autumn storms.

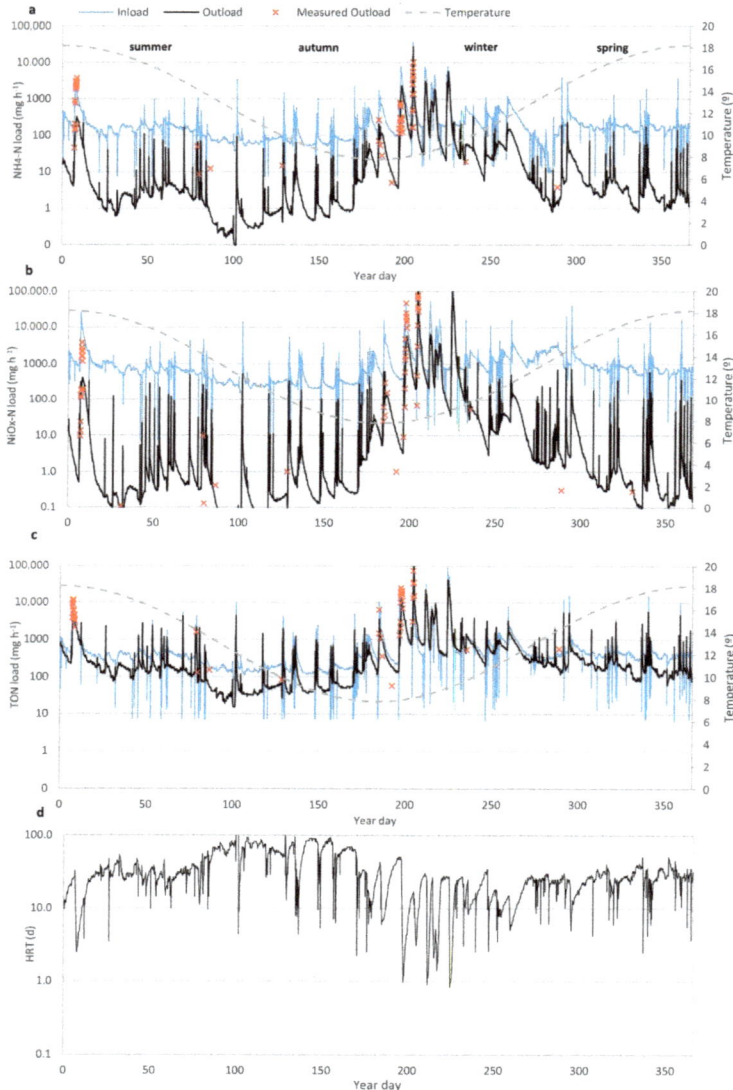

Figure 7. Time series for 2012 showing modelled and measured inflow and outflow NH4-N (**a**), NO_x-N (**b**) and TON (**c**) loads at UW and LW; and (**d**) modelled hydraulic retention times (HRT).

Table 5. Estimated NH_4-N, NO_x-N, TON and TN daily mass loading and removal rates, and wetland performance for different flow conditions.

	Flow-Type	Daily Mass Loading Rate ($mg/m^2/day$)	Daily Mass Removal Rate ($mg/m^2/day$)	Performance (%)
NH_4-N	Winter/Spring storms	49	21	44
	Summer/Autumn storms	6	5	90
	Baseflow	4	1	78
	All	5	4	73
NO_3-N	Winter/Spring storms	730	492	67
	Summer/Autumn storms	48	47	97
	Baseflow	21	16	80
	All	33	25	76
TON	Winter/Spring storms	328	114	35
	Summer/Autumn storms	38	17	44
	Baseflow	11	2.0	20
	All	16	4	26
TN	Winter/Spring storms	1147	589	51
	Summer/Autumn storms	95	71	75
	Baseflow	41	16	40
	All	60	35	57

4. Discussion

Water quality monitoring showed that the concentrations of NH_4-N, NO_x-N, Total Dissolved Nitrogen and Total-N were consistently lower at the outlet of the wetland regardless of flow conditions or seasonality, and, during the winter storms, all pollutants had significantly lower values. This was a strong indication that the wetland is very efficient at attenuating the poor water quality entering from upstream. Although the water quality measurements at the upper and lower wetland weirs indicated highest concentrations for TON in the surface water, the N entering in groundwater was mainly in the form of NO_x-N, with relatively lower TON concentrations. The dynamic hydrological sub-model calibrated with flow data from upper and lower weirs and local climatological data provided a rational approach to estimate relative inflows to the wetland from different pathways. It showed the wetland was almost 95% groundwater fed. Linking this information with median concentrations of N species measured in the different inflows to the wetland showed that NO_x-N was the dominant form of N entering (estimated as ~61% of load). The first order removal model estimated annual wetland net removal efficiency to be approximately 76% for NO_x-N, 73% for NH_4-N and 26% for TON. These forms of N are of course subject to transformations between different organic and inorganic forms via microbial processes (nitrification, denitrification, anammox and DNRA), assimilation by plants and microbes, adsorption to particulate phases (particularly NH_4-N), and settling and storage of particulate TON in the wetland. We did not have sufficient data about relative changes in N species or balance between different processes to be able to usefully infer sequential processing rates (e.g., [39]) in these large and highly dynamic systems.

The apparent treatment efficiency (percentage removal) of the headwater wetland was high for NO_x-N and TN compared with that of surface-flow constructed wetlands receiving tile drainage [3,33]. The wetland influent N loading rates in the present study were relatively low (TN ~60, NO_x-N 33 mg m^{-2} day^{-1}) compared to rates in excess of 400 mg m^{-2} day^{-1} measured in the constructed wetlands receiving tile drainage [3]. However, the best fit k_{20} reaction rate coefficient for NO_x-N was also very high compared to those measured in field scale constructed wetlands [4,33].

As with many headwater pastoral headwater wetlands in New Zealand, our study wetland is likely to have formed since European settlement of New Zealand. Since the catchment was cleared of its indigenous forest land cover during the 19th and early 20th centuries, clay-rich soils have eroded from the steep slopes and accumulated in the bottom of the gully to form a wetland swale stabilised by the dense sward of amphibious glaucous sweetgrass (*Glyceria declinata*). Similar features have been described from south-east Australia [40]. Our study wetland was strongly was by groundwater. However, the wetland was also losing water occasionally to groundwater. We detected slight spatial declining trend along the wetland in nitrate-N and Total-N concentrations, which has been also detected by other studies. Xu et al. [9] also detected decline in Total-N and nitrate-N concentrations along the wetland but not in the concentrations of ammonium, which may be removed by transient sorption, nitrification, anammox and plant uptake, but also generated through mineralization of organic N, and dissimilatory nitrate reduction to Ammonium [4]. Ackerman et al. [41], Maxwell et al. [42], and Brauer et al. [43] detected significant pollutant reduction along the groundwater flow paths associated with seepage from a wetland. They illustrate that subsurface flux away from wetlands provides another mechanism for N removal. Shallow groundwater infiltrating though the base and sides of the wetland would pass through saturated layers of anoxic sedimentary and organic material and plant detritus with high denitrification potential [7,35,44]. High denitrification rates are common in wetlands and riparian areas that are rich with organic material. For example, Whitmire and Hamilton [45] using push–pull techniques found that nitrate added to groundwater and re-injected into wetland sediments disappeared rapidly relative to conservative tracers without any lag time, and was depleted to below detection limits (10–15 µg N L^{-1}) within 5 to 20 h. In a ^{15}N denitrification study in a similar natural riparian wetland intercepting shallow groundwater, Burns and Nguyen [46] measured >90% removal of injected ^{15}NO$_3$-N along a 100 cm flow path through the organic wetland soil, with essentially all of the ^{15}NO$_3$-N- removed within 30 cm. Extensive ^{15}N tracer study performed on small headwater agricultural streams in the USA by Peterson [21] showed that NH$_4$-N and NO$_x$-N entering the streams was removed within tens to hundreds of meters and in one of the streams the denitrification rate was 0.045 µg N m^{-2} s^{-1} [47]. Denitrification consisted almost entirely of N$_2$ production (99%), with very little N$_2$O production occurring.

To describe the temperature response of denitrification, we used Arrhenius equation, which is most commonly used in this case and where the temperature has a stronger effect on the denitrification at lower temperatures [4,47]. Several authors have reported that denitrification has a "breakpoint" temperature, below which denitrification rates decrease more rapidly [9,48,49]. The exact temperature of the breakpoint is not known, but, in most of the studies, it varies usually between 10 and 12 °C [50]. However, Holtan-Hartwig et al. [51] found significant decrease in denitrification rates below 0 degrees and reported major deviations from the regular Arrhenius response at near freezing temperatures. Our results showed lower removal efficiency between 8 and 10 degrees, which coincided with high flows from surface runoff, which also had high loads of pollutants. Some studies have showed that wetlands receiving very variable flows show reduced nitrate removal performance [3,52], whereas Hernandez and Mitsch [53] found in the opposite that denitrification rates in the high marsh zone were significantly higher under flood pulsing (778 ± 92 µg N m^{-2} h^{-1}) than under steady flow (328 ± 63 µg N m^{-2} h^{-1}), but, in the low marsh and edge zones, flood pulses did not affect denitrification. Kadlec [4] also argues that, if wetland receives pollutants in pulses and the water interval between flows is sufficiently long, then the wetland will contain very low nitrate concentrations when another pulse arrives due to inter-event treatment. In the present study, removal efficiency declined during winter period, which coincided with low temperatures. Plants have been found to improve the wetland removal efficiency during the cold season. Yates et al. [54] demonstrated that *Carex aquatilis* helped to enhance nitrogen removal at cold temperatures and high hydraulic load. In addition, extended HRT could improve TN removal at low temperatures [9].

High denitrification rates are likely to be stimulated by the *Glyceria* vegetation that dominated the wetland. Matheson et al. [8], investigating the fate of ^{15}N-NO$_3$ in wetland soil microcosms, found that, in the presence of *Glyceria*, denitrification was elevated compared to unplanted microcosms and was the principal mechanism of NO$_x$-N removal (61–63%). They attributed high denitrification rates and low Dissimilatory Nitrate Reduction to Ammonium (DNRA) to differences in soil redox state induced by the presence of the plant. During the growth season, plant and periphyton uptake of dissolved inorganic forms of N is also likely to have occurred. Subsequent senescence and decay of this plant material is likely to have generated autochthonous particulate and dissolved organic N that was exported from the wetland.

Organic N export from agricultural land is often overlooked as a significant pathway of diffuse N loss from pasture lands [55,56], which, in some instances, can be of similar magnitude to NO$_x$-N losses. The estimated load of TON exiting from our wetland was 6.5 kg, which equates to 1.3 kg ha^{-1}. This is lower than the median level reported by van Kessel [55] (4.0 kg N ha^{-1} year^{-1}), and compared to dissolved organic N leaching rates measured in lowland pastoral soils elsewhere in New Zealand of 28–117 kg N ha^{-1} year^{-1} [57].

TON, the second largest component of TN entering the wetland, passed through the wetland with 26% net reduction. Wetlands characteristically accumulate a large pool of organic C and N, and experience with constructed wetlands suggests incoming organic nitrogen is reduced to a non-zero background concentration, created by residuals and wetland return fluxes [35]. A significant proportion of the TON load entered the wetland via surface runoff during storms, suggesting that much may have been associated with particulates [20]. Wilcock [36] previously studied the narrow (1–3 m) riparian-planted wetland swale that starts ~65 m below the lower weir of the headwater wetland in the present study. Based on their first two years of data, it is likely that around 50–60% of the TON exiting the headwater wetland would have been particulate organic N, likely associated with fine particulate detritus from plant litter and sloughed biofilm flocs. The remainder of the TON exported is expected to be dissolved organic forms, a significant proportion of which is likely to be bioavailable [58,59]. Analysis of the mass loading and removal data of Wilcock [36] shows that around 80% of the DON was removed in the ~350 m length of relatively aerobic wetland swale downstream of our lower weir, suggesting that much of it was not recalcitrant. There are several processes that can lead to a decrease in the DON concentration as it moves through the soil profile. Dissolved organic N is used as a substrate by soil microbes [55]. Wymore et al. [60] hypothesize that under a high N scenario microbes may be more energy limited (i.e., low C:N ratio) and DON may serve primarily as an energy source. However, they also found strong underlying seasonal patterns in the response to NO$_3$$^-$ addition, indicating that the role of DON can switch between serving as a nutrient source to an energy source. The other process that can lead to a decrease in the DON concentration is the uptake of DON by plants. The uptake of organic N has been found to be widespread across different ecosystems and to consist of an important source of N, in particular in ecosystems where microbial biomass is prone to large seasonal fluctuations and contributes to the release of labile organic N [55,61].

For modelling of TON removal, we used a relatively low background concentration (20 mg/m^3) based on concentrations measured during very low flow periods and in the wetland downstream [36]. The effect of background concentration on modelling results was evident as most of the time wetland received TON levels only slightly higher than the background concentration and therefore the first order removal was applied to a very small proportion of TON, which resulted also in low removal efficiency. During the storms, TON in loads was considerably higher than background concentration and the removal efficiency was also higher. There also appears to be little or no temperature dependence of organic nitrogen *k*-values [35]. This was also evident in this wetland as summer and winter TON concentrations for lower weir did not differ significantly (Figure 4). For modeling, we used a low temperature coefficient (1.01, Table 2), which resulted in very small seasonal variation in TON removal.

NH$_4$-N constituted only 9.4% of the inflow and the overall reduction was up to 73%. Most of the NH$_4$-N likely originated from the mineralised urea in cattle urine patches. The amount of N under a urine patch can be equivalent to an application rate of 700 to 1200 kg N ha^{-1} [62], much higher than the demand of N for any agricultural crop [55]. NH$_4$-N is commonly the preferred form of N for uptake by wetland plants, and, therefore, a proportion may have been assimilated by plants within the wetland during the growth season.

It is interesting to note that cattle grazing of the study wetland had only a very limited immediate impact on water quality [29]. According to Hughes et al. [29], a measurable increase in pollutants (attributable to cattle generated disturbance) occurred only once out of 18 high-intensity grazing days. This occurred when a cow became entrapped in close proximity to the downstream monitoring weir on a day when wetland flow was elevated.

The nitrogen (especially NO$_x$-N and NH$_4$-N) removal efficiency of our wetland was likely reduced by variability of inflow rates because the wetland received high nutrient level pulses during the winter storms, which coincided with lower temperatures [33]. This may have been exacerbated by water flowing across the wetland surface and around the wetland edges during high flow events [22,44]. Although the removed load during the storms was on an average 10 times higher than during baseflow, the removal efficiency (percentage removal) was lower. Nutrient removal by the wetland could be further improved by increasing residence time in the wetland through construction of a series of check dams and/or outflow control structures, although these would be prone to damage and erosion during storm-flows. The alternative of buffering inflow variability would appear to be practically difficult in the steep terrain surrounding the wetland, without changing from pasture to shrub or forest.

5. Conclusions

In comparison to constructed wetlands receiving defined surface flows (e.g., wastewaters or tile drainage), estimating the nutrient removal effectiveness of natural wetlands receiving a variable mix of seepage and overland flows is challenging. The measured concentrations of nitrate-N, Dissolved Inorganic Nitrogen and Total-N were always lower at the outlet of the wetland regardless of flow conditions or seasonality, even during winter storms. This indicated that the wetland is an effective removal mechanism for N. Use of the simple dynamic model calibrated to continuous flow monitoring, meteorological data and periodic water quality sampling enabled the loads from different pathways to be partitioned and load reductions estimated. Despite the highly variable flow of the wetland, it showed very high net overall reduction of NO_x-N and NH_4-N. Seepage of shallow nitrate-rich groundwater in through the organic-rich, anaerobic base of the wetland is likely to have stimulated the high denitrification rates modelled. Net removal of TON was less effective, likely due to decay processes in the wetland ecosystem, which regenerate N from plant detritus and biofilms, resulting in a natural background concentration. The results show that small seepage wetlands in the headwaters of New Zealand streams can be very effective at removing nitrogen loads. As their nutrient removal efficiency per unit land area is high, farmers might be more willing to maintain these natural wetlands undrained and fence them to prevent stock access. The study helps to raise awareness of the functions and values of these small pastoral seepage wetlands located in the headwater areas of the catchments.

Acknowledgments: The authors thank Jonathan Armstrong for permitting this study to be carried out on his property and allowing unlimited access to the monitoring sites. George Timpany, George Payne, Pete Pattinson, Kerry Costley, James Sukias and Gareth van Assema (NIWA) undertook site setup and maintenance. Water quality samples were analysed by NIWA-Hamilton's Water Quality Laboratory. The authors also wish to acknowledge Bob Kadlec (Wetland Management Services, Chelsea, MI, USA) for original conception of the modelling approach we have elaborated on here, Sandy Elliot (NIWA) for his advice on the modelling, and Kit Rutherford for review and useful comments on the manuscript. The first author is thankful to NIWA for hosting her for two years and for funding from the Marie Skłodowska-Curie Actions individual fellowships offered by the Horizon 2020 Programme under Research Executive Agency grant agreement number 660391, the Ernst Jaakson memorial stipend by University of Tartu Foundation, and institutional grant no. IUT 2-16 funded by the Estonian Ministry of Education and Research. Involvement of the other authors was supported through the AgResearch-led Clean Water: Productive Land Programme (C10X1006) and NIWA's Strategic Science Investment Funding (Ministry of Business, Innovation and Employment). We thank all the anonymous reviewers for insightful comments on the paper, as these comments led us to an improvement of the work.

Author Contributions: Chris C. Tanner and Andrew O. Hughes conceived the study and designed the monitoring programme; Chris C. Palliser, Evelyn Uuemaa and Chris C. Tanner developed and tested the model; Evelyn Uuemaa, Chris C. Palliser, Andrew O. Hughes and Chris C. Tanner wrote the paper.

Conflicts of Interest: The authors declare no conflict of interest.

References

1. Passeport, E.; Vidon, P.; Forshay, K.J.; Harris, L.; Kaushal, S.S.; Kellogg, D.Q.; Lazar, J.; Mayer, P.; Stander, E.K. Ecological Engineering Practices for the Reduction of Excess Nitrogen in Human-Influenced Landscapes: A Guide for Watershed Managers. *Environ. Manag.* **2013**, *51*, 392–413. [CrossRef] [PubMed]

2. Wang, J.L.; Yang, Y.S. An approach to catchment-scale groundwater nitrate risk assessment from diffuse agricultural sources: A case study in the Upper Bann, Northern Ireland. *Hydrol. Process.* **2008**, *22*, 4274–4286. [CrossRef]

3. Tanner, C.; Sukias, J. Multiyear nutrient removal performance of three constructed wetlands intercepting tile drain flows from grazed pastures. *J. Environ. Qual.* **2011**, *40*, 620–633. [CrossRef] [PubMed]

4. Kadlec, R.H. Constructed marshes for nitrate removal. *Crit. Rev. Environ. Sci. Technol.* **2012**, *42*, 934–1005. [CrossRef]

5. Baker, L.A. Design considerations and applications for wetland treatment of high-nitrate waters. *Water Sci. Technol.* **1998**, *38*, 389–395.

6. Fisher, J.; Acreman, M.C. Wetland nutrient removal: A review of the Wetland nutrient removal: A review of the evidence. *Hydrol. Earth Syst. Sci. Discuss.* **2004**, *8*, 673–685. [CrossRef]

7. Davidsson, T.E.; Stahl, M. The influence of organic carbon on nitrogen transformations in five wetland soils. *Soil Sci. Soc. Am. J.* **2000**, *64*, 1129–1136. [CrossRef]

8. Matheson, F.E.; Nguyen, M.L.; Cooper, A.B.; Burt, T.P.; Bull, D.C. Fate of 15N-nitrate in unplanted, planted and harvested riparian wetland soil microcosms. *Ecol. Eng.* **2002**, *19*, 249–264. [CrossRef]

9. Xu, J.H.; He, S.B.; Wu, S.Q.; Huang, J.C.; Zhou, W.L.; Chen, X.C. Effects of HRT and water temperature on nitrogen removal in autotrophic gravel filter. *Chemosphere* **2016**, *147*, 203–209. [CrossRef] [PubMed]

10. Ambus, P.; Christensen, S. Denitrification variability and control in a riparian fen irrigated with agricultural drainage water. *Soil Biol. Biochem.* **1993**, *25*, 915–923. [CrossRef]

11. Bernal, B.; Anderson, C.J.; Mitsch, W.J. Nitrogen Dynamics in Two Created Riparian Wetlands over Space and Time. *J. Hydrol. Eng.* **2017**, *22*. [CrossRef]

12. Zaman, M.; Nguyen, M.L.; Gold, A.J.; Groffman, P.M.; Kellogg, D.Q.; Wilcock, R.J. Nitrous oxide generation, denitrification, and nitrate removal in a seepage wetland intercepting surface and subsurface flows from a grazed dairy catchment. *Aust. J. Soil Res.* **2008**, *46*, 565–577. [CrossRef]

13. Matthew, C.; Horne, D.J.; Baker, R.D. Nitrogen loss: An emerging issue for the ongoing evolution of New Zealand dairy farming systems. *Nutr. Cycl. Agroecosyst.* **2010**, *88*, 289–298. [CrossRef]

14. Statistics New Zealand. Agriculture, Horticulture, and Forestry. 2016. Available online: http://www.stats. govt.nz/infoshare/ (accessed on 7 March 2017).

15. Parfitt, R.L.; Stevenson, B.A.; Dymond, J.R.; Schipper, L.A.; Baisden, W.T.; Ballantine, D.J. Nitrogen inputs and outputs for New Zealand from 1990 to 2010 at national and regional scales. *N. Z. J. Agric. Res.* **2012**, *55*, 241–262. [CrossRef]

16. Tanner, C.; Sukias, J.; Burger, D. Realising the value of remnant wetlands as farm attenuation assets. In *Moving Farm Systems to Improved Nutrient Attenuation*; Currie, L.D., Burkitt, L.L., Eds.; Occassional Report No. 28; Fertilizer and Lime Research Centre, Massey University: Palmerston North, New Zealand, 2015.

17. Ausseil, A.G.E.; Dymond, J.R.; Shepherd, J.D. Rapid mapping and prioritisation of wetland sites in the Manawatu-Wanganui region, New Zealand. *Environ. Manag.* **2007**, *39*, 316–325. [CrossRef] [PubMed]

18. Gluckman, P. *New Zealand's Fresh Waters: Values, State Trends and Human Impacts*; Office of the Prime Minister's Chief Science Advisor: Wellington, New Zealand, 2017. Available online: http://www.pmcsa.org.nz/wp-content/uploads/PMCSA-Freshwater-Report.pdf (accessed on 7 February 2018).

19. Organization for Economic Cooperation and Development (OECD). *Third Environmental Performance Review: New Zealand 2017*; Organization for Economic Cooperation and Development (OECD): Paris, France, 2017.

20. McKergow, L.A.; Rutherford, J.C.; Timpany, G.C. Livestock-Generated Nitrogen Exports from a Pastoral Wetland. *J. Environ. Qual.* **2012**, *41*, 1681–1689. [CrossRef] [PubMed]

21. Peterson, B.J. Control of Nitrogen Export from Watersheds by Headwater Streams. *Science* **2001**, *292*, 86–90. [CrossRef] [PubMed]

22. Jordan, T.E.; Whigham, D.F.; Hofmockel, K.H.; Pittek, M.A. Nutrient and Sediment Removal by a Restored Wetland Receiving Agricultural Runoff. *J. Environ. Qual.* **2003**, *32*, 1534–1547. [CrossRef] [PubMed]

23. Kovacic, D.A.; David, M.B.; Gentry, L.E.; Starks, K.M.; Cooke, R.A. Effectiveness of constructed wetlands in reducing nitrogen and phosphorus export from agricultural tile drainage. *J. Environ. Qual.* **2000**, *29*, 1262–1274. [CrossRef]

24. Borin, M.; Tocchetto, D. Five year water and nitrogen balance for a constructed surface flow wetland treating agricultural drainage waters. *Sci. Total Environ.* **2007**, *380*, 38–47. [CrossRef] [PubMed]

25. Knox, A.K.; Dahlgren, R.A.; Tate, K.W.; Atwill, E.R. Efficacy of natural wetlands to retain nutrient, sediment and microbial pollutants. *J. Environ. Qual.* **2008**, *37*, 1837–1846. [CrossRef] [PubMed]

26. O'Geen, A.T.; Budd, R.; Gan, J.; Maynard, J.J.; Parikh, S.J.; Dahlgren, R.A. Mitigating Nonpoint Source Pollution in Agriculture with Constructed and Restored Wetlands. In *Advances in Agronomy*; Academic Press: Cambridge, MA, USA, 2010; Volume 108, ISBN 9780123810311.

27. Wilcock, R.J.; Monaghan, R.M.; Quinn, J.M.; Campbell, A.M.; Thorrold, B.S.; Duncan, M.J.; McGowan, A.W.; Betteridge, K. Land-use impacts and water quality targets in the intensive dairying catchment of the Toenepi Stream, New Zealand. *N. Z. J. Mar. Freshw. Res.* **2006**, *40*, 123–140. [CrossRef]

28. Müller, K.; Srinivasan, M.S.; Trolove, M.; McDowell, R.W. Identifying and linking source areas of flow and P transport in dairy-grazed headwater catchments, North Island, New Zealand. *Hydrol. Process.* **2010**, *24*, 3689–3705. [CrossRef]

29. Hughes, A.; Tanner, C.; McKergow, L.; Sukias, J. Unrestricted dairy cattle grazing of a pastoral headwater wetland and its effect on water quality. *Agric. Water Manag.* **2016**, *165*, 72–81. [CrossRef]

30. Porteous, A.; Mullan, B. *The 2012-13 Drought: An Assessment and Historical Perspective*; MPI Technical Paper No: 2012/18; Prepared for the Ministry for Primary Industries; National Institute of Water and Atmospheric Research: Wellington, New Zealand, 2013.

31. Allen, R.G.; Pereira, L.S.; Raes, D.; Smith, M. *Crop Evapotranspiration: Guidelines for Computing Crop Water Requirements*; Food and Agriculture Organization of the United Nations (FAO): Rome, Italy, 1998; p. 300, ISBN 9251042195.

32. StatSoft Inc StatSoft. Available online: http://www.statsoft.com/ (accessed on 1 November 2017).

33. Kadlec, R.H. Nitrate dynamics in event-driven wetlands. *Ecol. Eng.* **2010**, *36*, 503–516. [CrossRef]

34. French, R.H. *Open-Channel Hydraulics*; McGraw-Hill: New York, NY, USA, 1985.

35. Kadlec, R.H.; Wallace, S.D. *Treatment Wetlands*, 2nd ed.; CRC Press: Boca Raton, FL, USA, 2009; ISBN 9781566705264.

36. Wilcock, R.J.; Müller, K.; Van Assema, G.B.; Bellingham, M.A.; Ovenden, R. Attenuation of nitrogen, phosphorus and *E. coli* inputs from pasture runoff to surface waters by a farm wetland: The importance of wetland shape and residence time. *Water Air Soil Pollut.* **2012**, *223*, 499–509. [CrossRef]

37. Wells, N.S.; Baisden, W.T.; Horton, T.; Clough, T.J. Spatial and temporal variations in nitrogen export from a New Zealand pastoral catchment revealed by stream water nitrate isotopic composition. *Water Resour. Res.* **2016**, *52*, 2840–2854. [CrossRef]

38. Parfitt, R.L.; Baisden, W.T.; Schipper, L.A.; Mackay, A.D. Nitrogen inputs and outputs for new zealand at national and regional scales: Past, present and future scenarios. *J. R. Soc. N. Z.* **2008**, *38*, 71–87. [CrossRef]

39. Tanner, C.; Kadlec, R.; Gibbs, M.; Sukias, J.; Nguyen, M. Nitrogen processing gradients in subsurface-flow treatment wetlands—Influence of wastewater characteristics. *Ecol. Eng.* **2002**, *18*, 499–520. [CrossRef]

40. Zierholz, C.; Prosser, I.P.; Fogarty, P.J.; Rustomji, P. In-stream wetlands and their significance for channel filling and the catchment sediment budget, Jugiong Creek, New South Wales. *Geomorphology* **2001**, *38*, 221–235. [CrossRef]

41. Ackerman, J.R.; Peterson, E.W.; Van der Hoven, S.; Perry, W.L. Quantifying nutrient removal from groundwater seepage out of constructed wetlands receiving treated wastewater effluent. *Environ. Earth Sci.* **2015**, *74*, 1633–1645. [CrossRef]

42. Maxwell, E.; Peterson, E.W.; O'Reilly, C.M. Enhanced Nitrate Reduction within a Constructed Wetland System: Nitrate Removal within Groundwater Flow. *Wetlands* **2017**, *37*, 413–422. [CrossRef]

43. Brauer, N.; Maynard, J.J.; Dahlgren, R.A.; O'Geen, A.T. Fate of nitrate in seepage from a restored wetland receiving agricultural tailwater. *Ecol. Eng.* **2015**, *81*, 207–217. [CrossRef]

44. Rutherford, J.C.; Nguyen, M.L. Nitrate removal in riparian wetlands: Interactions between surface flow and soils. *J. Environ. Qual.* **2004**, *33*, 1133–1143. [CrossRef] [PubMed]

45. Whitmire, S.L.; Hamilton, S.K. Rapid Removal of Nitrate and Sulfate in Freshwater Wetland Sediments. *J. Environ. Qual.* **2005**, *34*, 2062–2071. [CrossRef] [PubMed]

46. Burns, D.A.; Nguyen, L. Nitrate movement and removal along a shallow groundwater flow path in a riparian wetland within a sheep-grazed pastoral catchment: Results of a tracer study. *N. Z. J. Mar. Freshw. Res.* **2002**, *36*, 371–385. [CrossRef]

47. Mulholland, P.J.; Valett, H.M.; Webster, J.R.; Thomas, S.A.; Cooper, L.W.; Hamilton, S.K.; Peterson, B.J. Stream denitrification and total nitrate uptake rates measured using a field ^{15}N tracer addition approach. *Limnol. Oceanogr.* **2004**, *49*, 809–820. [CrossRef]

48. Malhi, S.S.; McGill, W.B.; Nyborg, M. Nitrate losses in soils: Effect of temperature, moisture and substrate concentration. *Soil Biol. Biochem.* **1990**, *22*, 733–737. [CrossRef]

49. Addy, K.; Gold, A.; Nowicki, B.; McKenna, J.; Stolt, M.; Groffman, P. Denitrification capacity in a subterranean estuary below a Rhode Island fringing salt marsh. *Estuaries* **2005**, *28*, 896–908. [CrossRef]

50. Schaefer, S.C.; Alber, M. Temperature controls a latitudinal gradient in the proportion of watershed nitrogen exported to coastal ecosystems. *Biogeochemistry* **2007**, *85*, 333–346. [CrossRef]

51. Holtan-Hartwig, L.; Dörsch, P.; Bakken, L.R. Low temperature control of soil denitrifying communities: Kinetics of N_2O production and reduction. *Soil Biol. Biochem.* **2002**, *34*, 1797–1806. [CrossRef]

52. Tanner, C.; Kadlec, R. Influence of hydrological regime on wetland attenuation of diffuse agricultural nitrate losses. *Ecol. Eng.* **2013**, *56*, 79–88. [CrossRef]

53. Hernandez, M.E.; Mitsch, W.J. Denitrification in created riverine wetlands: Influence of hydrology and season. *Ecol. Eng.* **2007**, *30*, 78–88. [CrossRef]

54. Yates, C.N.; Varickanickal, J.; Cousins, S.; Wootton, B. Testing the ability to enhance nitrogen removal at cold temperatures with *C. aquatilis* in a horizontal subsurface flow wetland system. *Ecol. Eng.* **2016**, *94*, 344–351. [CrossRef]

55. Van Kessel, C.; Clough, T.; van Groenigen, J.W. Dissolved Organic Nitrogen: An Overlooked Pathway of Nitrogen Loss from Agricultural Systems? *J. Environ. Qual.* **2009**, *38*, 393–401. [CrossRef] [PubMed]

56. Vogt, E.; Braban, C.F.; Dragosits, U.; Durand, P.; Sutton, M.A.; Theobald, M.R.; Rees, R.M.; McDonald, C.; Murray, S.; Billett, M.F. Catchment land use effects on fluxes and concentrations of organic and inorganic nitrogen in streams. *Agric. Ecosyst. Environ.* **2015**, *199*, 320–332. [CrossRef]

57. Ghani, A.; Müller, K.; Dodd, M.; Mackay, A. Dissolved organic matter leaching in some contrasting New Zealand pasture soils. *Eur. J. Soil Sci.* **2010**, *61*, 525–538. [CrossRef]

58. Pellerin, B.A.; Wollheim, W.M.; Hopkinson, C.S.; McDowell, W.H.; Williams, M.R.; Vörösmarty, C.J.; Daley, M.L. Role of wetlands and developed land use on dissolved organic nitrogen concentrations and DON/TDN in northeastern U.S. rivers and streams. *Limnol. Oceanogr.* **2004**, *49*, 910–918. [CrossRef]

59. Seitzinger, S.P.; Sanders, R.W.; Styles, R. Bioavailability of DON from natural and anthropogenic sources to estuarine plankton. *Limnol. Oceanogr.* **2002**, *47*, 353–366. [CrossRef]

60. Wymore, A.S.; Rodríguez-Cardona, B.; McDowell, W.H. Direct response of dissolved organic nitrogen to nitrate availability in headwater streams. *Biogeochemistry* **2015**, *126*, 1–10. [CrossRef]

61. Lipson, D.; Näsholm, T. The unexpected versatility of plants: Organic nitrogen use and availability in terrestrial ecosystems. *Oecologia* **2001**, *128*, 305–316. [CrossRef] [PubMed]

62. Jarvis, S.C.; Scholefi, D.; Pain, B. Nitrogen cycling in grazing systems. In *Nitrogen Fertilization in the Environment*; Bacon, P.E., Ed.; Marcel Dekker: New York, NY, USA, 1995.

water

MDPI

Article

Applicability of Constructed Wetlands for Water Quality Improvement in a Tea Estate Catchment: The Pussellawa Case Study

G. M. P. R. Weerakoon [1], K. B. S. N. Jinadasa [1,*], G. B. B. Herath [1], M. I. M. Mowjood [2] and W. J. Ng [3]

[1] Department of Civil Engineering, University of Peradeniya, Peradeniya 20400, Sri Lanka; prabhaw@pdn.ac.lk (G.M.P.R.W.); gemunuh@pdn.ac.lk (G.B.B.H.)

[2] Department of Agricultural Engineering, University of Peradeniya, Peradeniya 20400, Sri Lanka; mmowjood@pdn.ac.lk

[3] Environmental Bio-innovations Group (EBiG), Nanyang Technological University, Singapore 639798, Singapore; wjng@ntu.edu.sg

* Correspondence: shamj@pdn.ac.lk; Tel.: +94-812-239-3571

Received: 16 February 2018; Accepted: 12 March 2018; Published: 16 March 2018

Abstract: Water in agricultural catchments is prone to pollution from agricultural runoff containing nutrients and pesticides, and contamination from the human population working and residing therein. This study examined the quality of water in a drainage stream which runs through a congested network of 'line houses' (low-income housing, typically found arranged in straight 'lines' on estates) in the tea estate catchment area of Pussellawa in central Sri Lanka. The study evaluated the applicability of vertical subsurface flow (VSSF) and horizontal subsurface flow (HSSF) constructed wetlands for water polishing, as the residents use the stream water for various domestic purposes with no treatment other than possibly boiling. Water flow in the stream can vary significantly over time, and so investigations were conducted at various flow conditions to identify the hydraulic loading rate (HLR) bandwidth for wetland polishing applications. Two wetland models of 8 m × 1 m × 0.6 m (length × width × depth) were constructed and arranged as VSSF and HSSF units. Stream water was diverted to these units at HLRs of 3.3, 4, 5, 10, 20, and 40 cm/day. Results showed that both VSSF and HSSF wetland units were capable of substantially reducing five-day biochemical oxygen demand (BOD$_5$), total suspended solids (TSS), fecal coliform (FC), total coliform (TC), ammonia nitrogen (NH$_4{}^+$-N), and nitrate nitrogen (NO$_3{}^-$-N) up to 20 cm/day HLR, with removal efficiencies of more than 64%, 60%, 90%, 93%, 70%, and 59% for BOD$_5$, TSS, FC, TC, NH$_4{}^+$-N, and NO$_3{}^-$-N, respectively, in the VSSF wetland unit; and more than 66%, 62%, 91%, 90%, 53%, and 77% for BOD$_5$, TSS, FC, TC, NH$_4{}^+$-N, and NO$_3{}^-$-N, respectively, in the HSSF wetland unit.

Keywords: agriculture catchment; constructed wetlands; contaminated drainage stream; hydraulic loading rate; pollutant removal

---

1. Introduction

The pollution of water resources is a growing problem around the globe, due to inadequately treated wastewater discharges from ever-increasing populations, industrial and economic development, and agricultural interventions, resulting in the decline of fresh water resources suitable for human consumption [1]. This may affect human health as well as lead to various socioeconomic conflicts among groups of people. It has been estimated that about 80% of diseases and over one-third of deaths in the developing world are caused by the ingestion of contaminated water [2]. According to the World Health Organization (WHO), diarrheal diseases are the leading cause of illness in the developing world, and the second-leading cause of death among children under five years of age [3]. Thus, effective control of the contamination of natural water bodies is necessary [4]. However, wastewater treatment is far from satisfactory in developing countries [5,6], especially in semiurban and rural areas, due to constrained economic conditions and the lack of information. Therefore, some contamination of waterbodies can be expected, and hence development of appropriate and affordable water and wastewater treatment technologies is necessary in these contexts [7].

In Sri Lanka, many drainage streams in agriculture-based catchments, e.g., tea estates, are vulnerable to pollution, resulting in health issues among users. Agriculture-based communities are typically clustered around streams, and often, because of their economic situation, do not pay adequate attention to sanitation and hygiene practices. In addition to affordability constraints, there are also social issues, such as low education levels and traditional habits, in waste disposal. For example, such residents do locate their cesspits at stream banks, and so risk polluting the water sources. The problems can be further compounded by space constraints, water scarcity, a shallow water table, and rocky ground conditions [8]. Rajapaksha [9] investigated fecal pollution in vulnerable small streams in the Pussella-Oya catchment following the severe hepatitis A outbreak in May 2007 at the downstream location of Gampola, Sri Lanka [10], and this data, as presented in Table S1 in Supplementary Materials, would indicate seemingly good quality water in terms of some water quality parameters, but may not be so in terms of others, such as those in the microbiological category.

While there are many techniques available to treat and improve the quality of water [11], the challenge in this paper's context is to identify a simple and economical method which requires little attention from the users, and yet be able to address water quality upgrading needs, e.g., microbiological needs, sufficiently. The creation of constructed wetlands (CWs) for wastewater treatment was thought to be a possible candidate and alternative to conventional systems [12,13]. CWs utilize wetland plants, soils, and associated microbial assemblages to remove pollutants [14–16]. They have been used in many parts of the world for wastewater treatment [5,17,18], but there have been few reports on their application in water polishing. Of concern would be the treatment efficiencies of CWs, depending on various factors such as influent pollutant characteristics, hydraulic loading rate (HLR), climatic variation, and the required effluent characteristics [19].

In general, CWs require larger land areas compared to conventional methods [20]. This is because CWs require longer hydraulic retention time (HRT) or lower HLR [21] for substantial reduction of pollutants. Metcalf and Eddy [22] reported that pollutant removal in CWs is more efficient at 4–15 days HRT. However, since the stream flow and pollutant loads in these agriculture-based drainage streams do fluctuate over time and with the latter typically low, there is a need to know how CWs would perform when faced with high hydraulic loading rates (i.e., possibly shorter HRTs) but otherwise relatively low loads in relation to specific pollutant measures. The objectives of this study included identifying possible causes of human excreta pollution in a selected agriculture-based drainage stream in a congested tea estate human settlement, and to investigate the treatment performance of vertical subsurface flow (VSSF) and horizontal subsurface flow (HSSF) CWs under various HLRs to identify the applicable HLR bandwidth of CWs used to polish stream water.

2. Materials and Methods

2.1. Selection of the Study Site

Following the hepatitis A outbreak in Gampola, Sri Lanka in 2007 [10], and evidence of fecal pollution in the streams in the catchment as reported by Rajapaksha [9], a stream running through a cluster of line houses at a tea estate in Pussellawa (7°06′40.5″ E, 80°38′14.5″ N) was selected for this study. To identify the level of pollution, water quality examinations, including for BOD_5, FC, TC, NH_4^+-N, and NO_3^--N, were carried out over a one-month period at one-week intervals. A map of the study area is shown in Figure 1, and photographs of the environmental condition at a number of locations in the area are shown in Figure 2.

Figure 1. Map of the study area showing congested housing plots. (https://www.google.lk/maps/place/Tea+Estate,+Pussellawa).

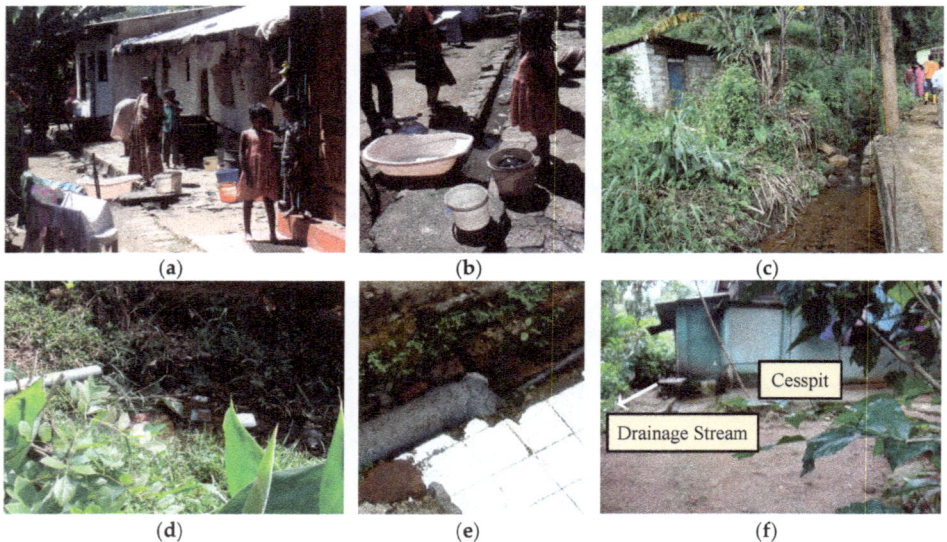

(a)   (b)   (c)

(d)   (e)   (f)

Figure 2. Photographs of the environmental conditions at locations in the study area. (**a**) A congested row of line houses. (**b**) A polluted drain which flows to the stream. (**c**) A pit latrine close to the drain. (**d**) Wastewater disposal to the drain through a pipe line. (**e**) Septic tank effluent directed to the stream. (**f**) A cesspit close to a drainage stream.

2.2. Identification of Sources of Stream Pollution

In order to identify the possible causes of stream pollution in the selected community, a questionnaire survey was conducted on the 74 households located along the stream. The data collected included information such as family size, age groups, sources of drinking water, available sanitation facilities, the distance between the sanitation facility and the stream, personal hygiene habits, gray water disposal methods, and waterborne disease history. The questionnaire is presented in Supplementary Materials. Data obtained from the survey was then analyzed to determine the routes of stream pollution.

2.3. Use of Constructed Wetlands for Stream Water Quality Improvement

Two units of subsurface flow CWs of size 8.0 m × 1.0 m × 0.6 m (length × width × height) were constructed using brick masonry and cement mortar, close to the selected stream as illustrated in Figure 3a. One was prepared as a HSSF wetland unit, while the other was prepared as a VSSF wetland unit, according to Figure 3b,c, respectively, using 10–20 mm gravel as the wetland media. To facilitate the easy distribution and collection of water in each unit, the inlet and outlet zones of the HSSF wetland unit and the drain field of the VSSF wetland unit were filled with 30–50 mm-sized gravel. In addition, each wetland unit had a surface layer of 10 cm-deep soil (<5 mm particle size) to support the vegetation. The two wetland units were planted with approximately 30 cm-high *Typha angustifolia* (Narrow-leaved Cattail) rhizomes, with each containing at least two nodes. The plantings were made 30 cm apart to achieve a plant density of 4 plants/m^2. The wetland units were then kept wet for four weeks for the plants to grow. Subsequently, a part of the stream water was diverted to a constant head tank, and applied to the wetland units at predetermined flow rates. This arrangement is shown in Figure 4.

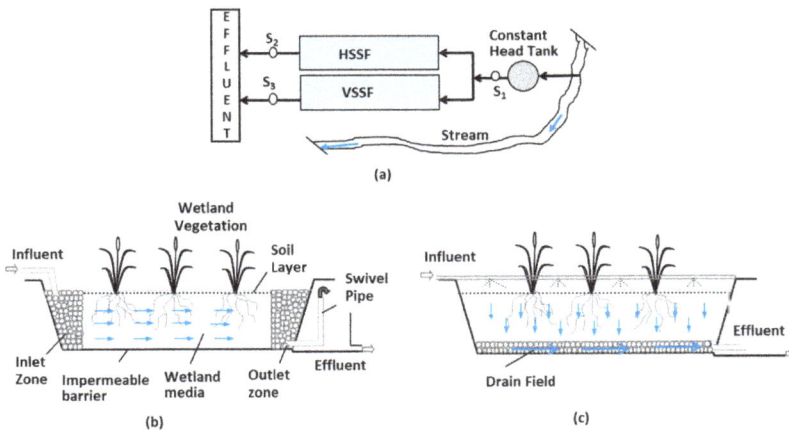

Figure 3. (**a**) Wetland arrangement: S_1, S_2 and S_3 are sampling points. (**b**) Schematic diagram of a horizontal subsurface flow (HSSF) wetland system. (**c**) Schematic diagram of a vertical subsurface flow (VSSF) wetland system.

Figure 4. Arrangement of wetland units in the field: (**a**) just after planting; and (**b**) mature wetland units.

2.4. Water Application

The performance of the VSSF and HSSF wetland units was investigated by diverting stream water at various HLRs in two phases as described below, according to the sequence shown in Figure 5.

(i)   Fed with stream water at three HLRs of 3.3, 4, and 5 cm/day

(ii) Fed with stream water spiked with a nitrogen source at HLRs of 10, 20, and 40 cm/day. This was to allow the investigation of situations where nitrogenous pollutants can be higher, as in during periods when the fertilization of tea plants with animal waste occurs.

The flow rate corresponding to each HLR was calculated using the wetland area, and flow was applied to each wetland unit using a flow control valve. The applied flow rates were monitored regularly to minimize fluctuations.

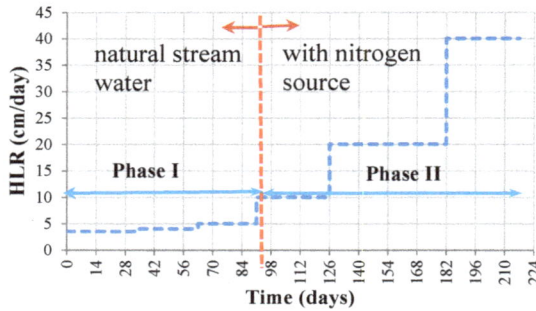

Figure 5. Sequence of water application in VSSF and HSSF wetland units. HLR: hydraulic loading rate.

2.5. Nitrogen Source

In order to evaluate the nitrogen removal capability of the VSSF and HSSF wetland units, cow dung was mixed with the stream water and stored in two 20 L cans. The mixture was then added to the wetland units' feed at a predetermined rate, using a pipe and valve arrangement as shown in Figure 6.

Figure 6. Arrangement for nitrogen (N) supplementation to the wetland units, H_{in}: Influent to the HSSF unit and V_{in}: Influent to the VSSF unit.

2.6. Sampling and Water Quality Analysis

In the first phase, influent and effluent samples were collected from the sampling points S_1, S_2, and S_3 (Figure 3). In the second phase, influent water samples were collected from H_{in} (Influent to the HSSF unit) and V_{in} (Influent to the VSSF unit) separately (Figure 6), while effluent samples were collected from S_2 and S_3. Samplings were conducted following a week of acclimatization after each flow adjustment. At least four samples were collected at each flow investigated at one-week intervals before switching into the next HLR level. Samples were collected in 500 mL cleaned plastic bottles, and sent to the Environmental Engineering laboratory, Faculty of Engineering, University of Peradeniya to measure the water quality parameters, such as five-day biochemical oxygen demand (BOD_5), total suspended solids (TSS), fecal coliform (FC), total coliform (TC), ammonia nitrogen (NH_4^+-N), and nitrate nitrogen (NO_3^--N), in accordance with APHA standard methods [23]. The removal efficiency (RE) of each parameter on each sampling occasion was calculated using Equation (1). The respective mass loading rates (MLR) and mass removal rates (MRR) in g-P/m^2·day (P = pollutant concentration) for BOD_5, TSS, NH_4^+-N, and NO_3^--N, as well as in cfu/m^2·day (cfu = colony forming units) for FC and TC, were calculated using Equations (2) and (3), respectively.

$$RE = \frac{C_i - C_o}{C_i} \times 100\% \qquad (1)$$

$$MLR = C_i \times HLR \qquad (2)$$

$$MRR = (C_i - C_o) \times HLR \qquad (3)$$

where, C_i = influent concentration and C_o = effluent concentration for the water quality parameter.

2.7. Statistical Analysis

In this study, statistical analysis was conducted using the "MINITAB 16" statistical software (16, Minitab Ltd., Coventry, UK). Normality of influent and effluent water quality characteristics was determined by performing the Anderson–Darling test. The significance in treatment differences between the VSSF and HSSF wetland systems subjected to various HLRs was evaluated using the one-way ANOVA test for normally distributed data and Mann–Whitney test for non-normal data at a 0.05 significance level. Correlation between MLR and MRR was identified using linear regression analysis.

3. Results

3.1. Characteristics of the Inhabitants

The total population in this community numbered 325 people, of which 53% were between 18 and 50 years of age, while 19% were between 5 and 17 years, 9% less than 5 years, and 19% more than 50 years of age. Their living conditions can be considered poor, with the majority living in line houses of either the single barracks (43%) or double barracks (49%) type. Only 8% owned their own homes.

Water availability could be summarized as: 84% of households had piped water supply, 14% used unprotected wells, and 2% used spring water for drinking and cooking. About 7% of the people had used stream water to wash kitchen utensils, bathing, and in their toilets. Even though a large portion of the group was supplied with piped water, the quality was aesthetically poor as sediments were present. Nonetheless, the incidence of water-borne diseases in this community was low.

In terms of sanitary facilities, 95% of homes had latrines, while 2% shared their neighbors' latrines. However, 3% defecated on open ground near the stream, and used water from the stream to wash after defecation. There were also six persons with walking difficulties in this community, two of whom disposed excreta into the stream. In addition, excreta of 31% of the children (<2 years old) were disposed of on the open ground or washed into the stream. Almost all latrines in this community were pit latrines/pour-flush latrines with cesspits, and 26% of them were very close to the stream (<15 m). All of these contribute to the pollution of stream water. In addition, 22% of households directed their gray water into drains which finally flow into the stream, 38% discharge it into longer drains (>20 m long) leading elsewhere, and only 40% had safe disposal.

3.2. Stream Water Characteristics

Average influent stream water and VSSF and HSSF wetland unit effluent characteristics operated with 3.3–5 cm/day HLRs are shown in Table 1, and the characteristics of stream water spiked with the nitrogen source at the influents and effluents of VSSF and HSSF wetland units operated with 10–40 cm/day HLRs are shown in Table 2. Since NH_4^+-N and NO_3^--N concentrations were very low in the natural stream water (less than 0.40 and 1.76 mg/L, respectively), they were not analyzed in the first phase of the study. It was observed that the average water quality had varied over the period, and this could be due to rain during the study period.

Statistical analysis showed that only BOD_5 and TSS in the influents and effluents of both VSSF and HSSF wetland units operated with natural stream water, as well as water spiked with a nitrogen source, were normally distributed ($p > 0.05$) while FC, TC, NH_4^+N, and NO_3^--N were not normally distributed ($p < 0.05$) at a 95% significance level. The average quality variation corresponding to different HLRs at the influent and effluents of VSSF and HSSF wetland units for BOD_5, TSS, FC, TC, NH_4^+N, and NO_3^--N are shown in Figure S1 in Supplementary Materials.

Table 1. Average characteristics of natural stream water at the influent and effluents of VSSF and HSSF wetland units.

Parameter	Influent	Effluents	
		VSSF	HSSF
BOD_5 (mg/L)	3.2 ± 0.6	0.7 ± 0.3	0.9 ± 0.4
TSS (mg/L)	157.8 ± 56	49.2 ± 22.7	42.7 ± 21.5
FC (cfu/100 mL)	793 ± 389	7 ± 5	2 ± 3
TC (cfu/100 mL)	1669 ± 853	26 ± 19	33 ± 23

Table 2. Average characteristics of stream water spiked with a nitrogen source at the influents and effluents of VSSF and HSSF wetland units.

Parameter	VSSF		HSSF	
	Influent	Effluent	Influent	Effluent
BOD_5 (mg/L)	9.7 ± 2.2	3.4 ± 0.7	9.3 ± 1.9	3.1 ± 1.0
TSS (mg/L)	236.8 ± 91.2	105.4 ± 63.0	245.6 ± 103.4	93.9 ± 54.3
FC (cfu/100 mL)	1672 ± 1411	226 ± 218	1653 ± 1324	188 ± 167
TC (cfu/100 mL)	3115 ± 2353	353 ± 410	3028 ± 1904	353 ± 286
NO_3^--N (mg/L)	14.9 ± 10.6	6.3 ± 3.9	15.7 ± 12.7	3.8 ± 2.9
NH_4^+-N (mg/L)	9.39 ± 3.34	2.25 ± 1.63	9.78 ± 3.18	4.57 ± 2.87

4. Discussion

4.1. Drainage Stream Water Characteristics

The selected drainage stream, which flowed through the human settlement, was more contaminated with fecal coliforms than with organic matter and nutrients (Table 1), exceeding the proposed ambient water quality standards for inland waters in Sri Lanka [24]. This was obviously a consequence of the inadequate sanitation arrangements. From the survey, it was noted that open defecation was still practiced, and was followed by cleaning at the stream. Excreta generated in homes were also sometimes disposed at the stream. Furthermore, people washed small children after defecation into the house drains, which then lead to the stream. The latrines and cesspits were old and located close to the stream. It was therefore possible for contamination from the pits to leak into the stream. However, BOD_5 concentration in the stream water was low, and this was likely due to dilution. The presence of fecal coliforms indicated a higher risk of pathogens being present in water, which may cause waterborne diseases.

4.2. Pollutant Removal

Figure 7 shows the variation in average removal efficiencies (REs) of BOD_5, TSS, FC, TC, NH_4^+-N, and NO_3^--N at various HLRs in the VSSF and HSSF wetland units. The VSSF and HSSF units had the same removal trend for all the measured water quality parameters: i.e., a decreasing RE with increasing HLR; except for NH_4^+-N, which showed an increase of RE corresponding to increasing HLRs. This was seemingly an unusual phenomenon and contradicted expectations, and will be investigated further.

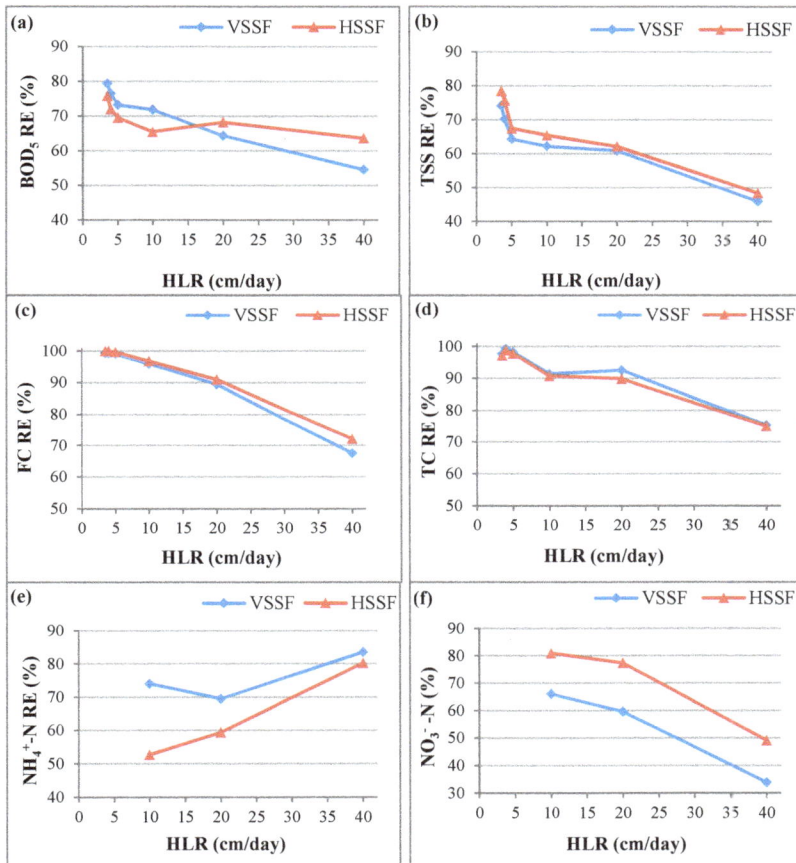

Figure 7. Variation of removal efficiencies (RE) of (**a**) BOD_5, (**b**) TSS, (**c**) FC, (**d**) TC, (**e**) NH_4^+-N, and (**f**) NO_3^--N in VSSF and HSSF wetland units versus HLR.

BOD_5 removal mechanisms in CWs include adsorption, sedimentation, filtration, and microbial degradation [25]. From Figure 7a, even though it was noted that there are slight differences in BOD_5 removal between the VSSF and HSSF wetland units, statistical analysis showed there was no significant treatment difference between the two systems ($p > 0.05$).

TSS removal in a wetland system is supported by physical processes such as filtration, sedimentation, and surface adhesion, followed by microbial assimilation within the substrate media [26]. Figure 7b shows that TSS removal efficiency in the VSSF and HSSF wetlands had been similar during the study period, though the HSSF unit had shown slightly higher removal.

Statistical analysis again indicated there was no significant treatment difference between the two wetland systems ($p > 0.05$).

Coliform removal in CWs is achieved through many physical, chemical, and biological processes, such as sedimentation, filtration, ultraviolet radiation, adsorption, oxidation, die-off due to toxins, natural die-off, and ingestion by nematodes and protozoa [27]. From Figure 7c,d, it can be noted that there was no treatment difference between the two wetland units with up to 10 cm/day HLR. Increasing HLRs beyond 10 cm/day caused a slight removal difference between the two systems for both FC and TC, but this was not statistically significant ($p > 0.05$).

In CWs, a variety of chemical and biological processes, such as ammonification, ammonia volatilization, nitrification, de-nitrification, microbial assimilation, plant uptake, and matrix adsorption, are involved in nitrogen removal [13,28]. Of these, nitrification and denitrification processes are considered the major nitrogen removal pathways [29]. For effective nitrification, the wetland bed has to be aerobic, and typically VSSF wetlands are considered aerobic. On the other hand, for effective denitrification, the wetland bed has to be anoxic, and HSSF wetlands are considered anoxic due to their waterlogged/saturated nature. Thus, VSSF wetlands are more prone to removing NH_4^+ through nitrification, and HSSF wetlands are more prone to removing NO_3^- through denitrification. From Figure 7e, it was noted that the VSSF unit was better in NH_4^+-N removal than the HSSF unit, though both wetland units had the same NH_4^+-N removal trend. Also, it was noted that NH_4^+-N removal increased with increasing HLRs. This could not be satisfactorily explained at this juncture, and the phenomenon will be investigated further. Statistical analysis showed there was no significant treatment difference between the two systems ($p > 0.05$) for NH_4^+-N removal. From Figure 7f, it was noted that both VSSF and HSSF units had a similar NO_3^--N removal trend: a decreasing removal efficiency corresponding to increasing HLRs. The HSSF unit removed more NO_3^--N than the VSSF wetland, and in this instance, statistical analysis showed that there was a significant treatment difference between the two wetland units ($p < 0.05$).

Furthermore, statistical analysis on the VSSF wetland unit showed no significant treatment difference in TC removal up to 5 cm/day HLR; FC and TSS removal up to 10 cm/day HLR; BOD_5 and NO_3^--N removal up to 20 cm/day HLR; and NH_4^+-N removal up to 40 cm/day HLR; with more than 99% of TC, 97% of FC, 62% of TSS, 64% of BOD_5, 60% of NO_3^--N, and 70% of NH_4^+-N removal. On the other hand, the HSSF wetland showed no significant treatment difference in FC and TC removal up to 10 cm/day HLR; NO_3^--N and TSS removal up to 20 cm/day; and HLR, BOD_5, and NH_4^+-N removal up to 40 cm/day HLR; with more than 97% of FC, 90% of TC, 77% of NO_3^--N, 62% of TSS, 63% of BOD_5, and 53% of NH_4^+-N removal. These results indicate that a 20 cm/day HLR would be the appropriate upper limit for the applied HLR for pollutant removal, with the HSSF configuration being more suitable for BOD_5, TSS, FC, and NO_3^--N removal. These results were consistent with laboratory-scale studies reported earlier by Weerakoon et al. [7] and Weerakoon et al. [21].

Table 3 presents the range of applied mass loading rates (MLRs) and mass removal rates (MRRs) obtained for the VSSF and HSSF wetland units for polishing stream water, and Table 4 presents the range of applied MLRs and MRRs obtained for the VSSF and HSSF units with spiked nitrogen, for BOD_5, TSS, FC, TC, NH_4^+-N, and NO_3^--N at different HLRs. MRRs show a positive response to increasing the HLR for both the VSSF and HSSF wetland units with all pollutants, except for NO_3^--N in the VSSF unit. This negative response for NO_3^--N MRRs in the VSSF unit is believed to be a result of the enhanced nitrification and incomplete denitrification under an oxic environment and lack of a suitable carbon source [30]. The regression analysis confirmed these observations, with a strong linear correlation between MLRs and MRRs for BOD_5, TSS, FC, TC, and NH_4^+-N, of over 0.9 in both wetland units; a moderate correlation for NO_3^--N ($R^2 = 0.743$) in the HSSF wetland unit; and a weak relationship for NO_3^--N in the VSSF wetland unit ($R^2 = 0.25$). On the other hand, REs were negatively impacted by increasing HLRs for all measurements, except for NH_4^+-N, which was positively impacted by increasing HLRs in both the VSSF and HSSF units, though such enhanced NH_4^+-N REs with increasing HLRs were not expected.

Table 3. Mass loading rates (MLRs) and mass removal rates (MRRs) for VSSF and HSSF wetland units polishing stream water.

Parameter	MLR	MRR	
		VSSF	HSSF
BOD$_5$ (g/m^2·day)	0.106–0.171	0.088–0.125	0.080–0.119
TSS (g/m^2·day)	3.815–9.9	2.826–6.362	2.992–6.675
FC (cfu/m^2·day)	$(1.65–4.95) \times 10^5$	$(1.64–4.90) \times 10^5$	$(1.64–4.93) \times 10^5$
TC (cfu/m^2·day)	$(3.78–10.29) \times 10^5$	$(3.70–10.29) \times 10^5$	$(3.68–10.06) \times 10^5$

Table 4. MLRs and MRRs for VSSF and HSSF wetland units polishing stream water spiked with a nitrogen source.

Parameter	VSSF		HSSF	
	MLR	MRR	MLR	MRR
BOD$_5$ (g/m^2·day)	1.196–3.088	0.865–1.67	1.092–3.289	0.712–2.094
TSS (g/m^2·day)	16.65–110.1	10.5–48.2	19.075–112.7	12.675–54.0
FC (cfu/m^2·day)	$(9.70–61.0) \times 10^5$	$(9.39–40.35) \times 10^5$	$(13.8–50.9) \times 10^5$	$(13.2–35.9) \times 10^5$
TC (cfu/m^2·day)	$(21.7–140.6) \times 10^5$	$(20.6–104.2) \times 10^5$	$(25.45–118.9) \times 10^5$	$(23.7–89.2) \times 10^5$
NH$_4^+$-N (g/m^2·day)	1.031–4.123	0.86–3.44	0.99–3.96	0.796–3.184
NO$_3^-$-N (g/m^2·day)	1.648–2.57	1.18–2.17	1.48–3.92	1.218–1.56

5. Conclusions

Water catchments within agricultural areas can be contaminated not only with agrochemicals but also by the human settlements within the catchment. A consequence of the latter would be the occurrence of fecal contamination. The latter can then lead to waterborne diseases, as residents of such settlements do make use of the water in streams for domestic purposes.

This study revealed that both the VSSF and HSSF wetlands planted with *Typha angustifolia* could improve water quality in such streams, and hence serve as a relatively low-cost and simple-to-operate water polishing device. Both the VSSF and HSSF wetland units reduced BOD$_5$, TSS, FC, TC, NH$_4^+$-N, and NO$_3^-$-N substantially within the 3.3–20 cm/day HLRs operating bandwidth, though the reduction of NO$_3^-$-N was lower in the VSSF wetland unit. This bandwidth of HLRs is important because in the application proposed, the wetland units would be more impacted by hydraulic load variations than by contaminant levels, as the stream water would not be heavily contaminated as in wastewater.

Supplementary Materials: The following are available online at http://www.mdpi.com/2073-4441/10/3/332/s1, Table S1: Water quality characteristics in several subcatchments in the Pussella-oya catchment during 2008; Table S2: Questionnaire for situation assessment close to the selected drainage stream in the selected tea estate, Pussellawa; Figure S1: The average water quality variation corresponding to different HLRs at the influent and effluents of VSSF and HSSF wetland units for (a) BOD$_5$, (b) TSS, (c) FC, (d) TC, (e) NH$_4^+$-N, and (f) NO$_3^-$-N.

Acknowledgments: The authors wish to acknowledge the CB project in the Postgraduate Institute of Agriculture, University of Peradeniya, Sri Lanka, and the UNESCO-IHE Institute for water education, Delft, The Netherlands (101-IRIS-07; IDD 90201/1/51) for providing funds to conduct this study.

Author Contributions: W.J.N., M.I.M.M., G.B.B.H., and K.B.S.N.J. conceived and designed the experiments; G.M.P.R.W. and M.I.M.M. performed the experiments; G.B.B.H. and W.J.N. analyzed the data; W.J.N. contributed reagents/materials/analysis tools; G.M.P.R.W. and K.B.S.N.J. wrote the paper.

Conflicts of Interest: The authors declare no conflict of interest.

References

1. Llorens, M.; Perez-Marin, A.B.; Aguilar, M.I.; Saez, J.; Ortuno, J.F.; Meseguer, V.F. Nitrogen transformation in two subsurface infiltration systems at pilot scale. *Ecol. Eng.* **2011**, *37*, 736–743. [CrossRef]
2. Amendola, M.; de Souza, A.L.; Roston, D.M. Numerical simulation of fecal coliform reduction at a constructed wetland. *Rev. Bras. Eng. Agríc. Ambient.* **2003**, *7*, 533–538. [CrossRef]
3. WHO. Fact Sheet, Diarrhoeal Diseases. Available online: www.who.int/mediacentre/factsheets/fs330/en/ (accessed on 4 February 2018).
4. Iasur-Kruth, L.; Hadar, Y.; Milstein, D.; Gasith, A.; Minz, D. Microbial population and activity in wetland microcosms constructed for improving treated municipal wastewater. *Environ. Microbiol.* **2009**, *59*, 700–709. [CrossRef]
5. Trang, N.T.D.; Konnerup, D.; Schierup, H.H.; Chiem, N.H.; Tuan, L.A.; Brix, H. Kinetics of pollutant removal from domestic wastewater in a tropical horizontal subsurface flow constructed wetland system: Effects of hydraulic loading rate. *Ecol. Eng.* **2010**, *36*, 527–535. [CrossRef]
6. Sato, T.; Qadir, M.; Yamamoto, S.; Endo, T.; Zahoor, A. Global, regional and country level need for data on wastewater generation, treatment and use. *Agric. Water Manag.* **2013**, *130*, 1–13. [CrossRef]
7. Weerakoon, G.M.P.R.; Jinadasa, K.B.S.N.; Herath, G.B.B.; Mowjood, M.I.M.; van Bruggen, J.J.A. Impact of hydraulic loading rate on pollutants removal in tropical horizontal subsurface flow constructed wetlands. *Ecol. Eng.* **2013**, *61*, 154–160. [CrossRef]
8. Gunawardana, I.P.P.; Rajapakshe, I.H.; Galagedara, L.W. Application of ecological sanitation technology/s to Pussella-Oy sub catchment—A concept note. In *Water Supply, Sanitation and Wastewater Management: Progress and Prospects Towards Clean and Healthy Society, Proceedings of the Symposium, Peradeniya, Sri Lanka, 23 June 2008*; Cap-Net Lanka: Peradeniya, Sri Lanka, 2009; pp. 99–106.
9. Rajapaksha, I.H. Investigation of Water Quality Variation and Pollution Sources in Pussella Oya Catchment. Master's Thesis, Postgraduate Institute of Agriculture, University of Peradeniya, Peradeniya, Sri Lanka, 2009.
10. Ministry of Health. *Annual Health Bulletin*; Ministry of Health: Colombo, Sri Lanka, 2007.
11. Mena, J.; Rodriguez, L.; Nunez, J.; Fernandez, F.J.; Villaenor, J. Design of horizontal and vertical sub-surface flow constructed wetlands treating industrial wastewater. *WIT Trans. Ecol. Environ.* **2008**, *111*, 555–564.
12. Noorvee, A. The Applicability of Hybrid Subsurface Flow Constructed Wetland Systems with Re-Circulation for Wastewater Treatment in Cold Climates. Ph.D. Thesis, University of Tartu, Tartu, Estonia, 2003.
13. Kadlec, R.H.; Wallace, S.D. *Treatment Wetlands*, 2nd ed.; CRC Press, Taylor and Francis Group: Boca Raton, FL, USA, 2009.
14. Farooqi, I.H.; Basheer, F.; Chaudhari, R.J. Constructed wetland systems (CWS) for wastewater treatment. In Proceedings of the Taal, the 12th World Lake Conference, Jaipur, India, 29 October–2 November 2007; pp. 1004–1009.
15. Truong, H.D.; Le, N.Q.; Nguyen, H.C.; Brix, H. Treatment of high-strength wastewater in tropical constructed wetlands planted with *Sesbania sesban*: Horizontal subsurface flow versus vertical down flow. *Ecol. Eng.* **2011**, *37*, 711–720.
16. Qasaimeh, A.; AlsSharie, H.; Masoud, T. A review on constructed wetlands components and heavy metal removal from wastewater. *J. Environ. Prot.* **2015**, *6*, 710–718. [CrossRef]
17. Kantawanichkul, S.; Polprasert, S.; Brix, H. Treatment of high-strength wastewater in tropical vertical flow constructed wetlands planted with *Typha angustifolia* and *Cyperus involucratus*. *Ecol. Eng.* **2009**, *35*, 238–247. [CrossRef]
18. Zhai, J.; Xiao, H.W.; Kujawa-Roeleveld, K.; He, Q.; Kerstens, S.M. Experimental study of a novel hybrid constructed wetland for water reuse and its application in southern China. *Water Sci. Tech.* **2011**, *64*, 2177–2184. [CrossRef] [PubMed]
19. Tanaka, N.; Jinadasa, K.B.S.N.; Werallagama, D.R.I.B.; Mowjood, M.I.M.; Ng, W.J. Constructed tropical wetlands with submergent-emergent plants for water quality improvement. *J. Environ. Sci. Health Part A* **2006**, *4*, 2221–2236. [CrossRef] [PubMed]
20. Deblina, G.; Brij, G. Effect of hydraulic retention time on the treatment of secondary effluent in a subsurface flow constructed wetland. *Ecol. Eng.* **2010**, *36*, 1044–1051.

21. Weerakoon, G.M.P.R.; Jinadasa, K.B.S.N.; Herath, G.B.B.; Mowjood, M.I.M.; Zhang, D.; Tan, S.K.; Jern, N.W. Performance of tropical vertical subsurface flow constructed wetlands at different hydraulic loading rates. *Clean Soil Air Water* **2016**, *44*, 1–11. [CrossRef]

22. Tchobanoglous, G.; Burton, F.L. *Wastewater Engineering: Treatment, Disposal and Reuse*; Metcalf and Eddy; McGrow Hill: New York, NY, USA, 1991; p. 1334.

23. American Public Health Association. *Standard Methods for the Examination of Water and Wastewater*, 20th ed.; American Public Health Association/American Water Works Association/Water Environment Federation: Washington, DC, USA, 2005.

24. Central Environmental Authority. *Proposed Ambient Water Quality Standards for Inland Waters Sri Lanka*; Interim Standard, Board Paper No. 361/3677/16; Central Environmental Authority: Colombo, Sri Lanka, 2016.

25. Merino-Solis, M.; Villegas, E.; de Anda, J.; Lopez-Lope, A. The effect of the hydraulic retention time on the performance of an ecological wastewater treatment system: An anaerobic filter with a constructed wetland. *Water* **2015**, *7*, 1149–1163. [CrossRef]

26. Dordio, A.; Palace, A.J.; Pinto, A.P. Wetlands: Water living filters? In *Wetlands: Ecology, Conservation and Restoration*; Russo, R.E., Ed.; Nova Science Publishers: Hauppauge, NY, USA, 2008.

27. Hoddinot, B.C. Horizontal Subsurface Flow Constructed Wetlands for on-Site Wastewater Treatment. Master's Thesis, Wright State University, Dayton, OH, USA, 2006.

28. Vymazal, J. Constructed wetlands for wastewater treatment: Five decades of experience. *Environ. Sci. Technol.* **2008**, *45*, 61–69. [CrossRef] [PubMed]

29. Lee, C.G.; Fletcher, T.D.; Sun, G. Nitrogen removal in constructed wetlands systems: Review. *Eng. Life Sci.* **2008**, *9*, 11–22. [CrossRef]

30. Lu, S.; Hu, H.; Sun, Y.; Yang, J. Effect of carbon source on the Denitrification in constructed wetlands. *J. Environ. Sci.* **2009**, *21*, 1036–1043. [CrossRef]

water

MDPI

Article

Long-Term Monitoring of a Surface Flow Constructed Wetland Treating Agricultural Drainage Water in Northern Italy

Stevo Lavrnić [1], Ilaria Braschi [1], Stefano Anconelli [2], Sonia Blasioli [1], Domenico Solimando [2], Paolo Mannini [2] and Attilio Toscano [1,*]

[1] Department of Agricultural and Food Sciences, Alma Mater Studiorum-University of Bologna, Viale Giuseppe Fanin 40-50, 40127 Bologna, Italy; stevo.lavrnic@unibo.it (S.L.); ilaria.braschi@unibo.it (I.B.); sonia.blasioli@unibo.it (S.B.)
[2] Consorzio di Bonifica Canale Emiliano Romagnolo, Via Ernesto Masi 8, 40137 Bologna, Italy; anconelli@consorziocer.it (S.A.); solimando@consorziocer.it (D.S.); mannini@consorziocer.it (P.M.)
* Correspondence: attilio.toscano@unibo.it; Tel.: +39-051-20-9-6179

Received: 28 February 2018; Accepted: 9 May 2018; Published: 16 May 2018

Abstract: Agricultural drainage water that has seeped into tile drainage systems can cause nitrogen and phosphorus pollution of the surface water bodies. Constructed wetlands (CWs) can help mitigate the effects of agricultural non-point sources of pollution and remove different pollutants from tile drainage water. In this study, hydrological and water quality data of a Northern Italian CW that has been treating agricultural drainage water since 2000 were considered to assess its ability to mitigate nitrogen and phosphorus pollution. The effects of such long-term operation on the nutrients and heavy metals that eventually accumulate in CW plants and sediments were also analysed. Since 2003, the CW has received different inflows with different nutrient loads due to several operation modes. However, on average, the outflow load has been 50% lower than the inflow one; thus, it can be said that the system has proved itself to be a viable option for tile drainage water treatment. It was found that the concentration of nitrogen and phosphorus in the plant tissues varied, whereas the nitrogen content of the soil increased more than 2.5 times. Heavy metals were found accumulated in the plant root systems and uniformly distributed throughout a 60 cm soil profile at levels suitable for private and public green areas, according to the Italian law

Keywords: agricultural drainage water; superficial flow constructed wetlands; nutrients; potentially toxic elements

1. Introduction

Constructed wetlands (CWs) are man-made systems in which processes occurring in natural wetlands are applied for water treatment [1]. Compared to conventional wastewater treatment plants, CWs are less costly, easier to operate, and require less maintenance [2]. In addition to their general ability to improve wastewater quality, CWs are multifunctional systems that can offer different ecosystem services [3], including supporting habitat and biodiversity functions, together with recreational and socioeconomic services, including flood and drought control, water retention, and erosion prevention [4–6]. These systems can also be used as a part of water reuse schemes [7,8] and can have positive effects on carbon balance [9]. CWs are usually characterized by high evapotranspiration rates due to the presence of vegetation. In addition, the biomass produced can later be exploited for energy production [10,11]. CWs have been successfully used for treatment of many different types of wastewater [3,12,13], including agricultural drainage water. In general, CWs are more cost-effective for reducing non-point source pollution than other measures [14].

Agricultural drainage water contains different substances that can have negative effects on surface water bodies, and therefore its treatment is important in order to reduce environmental pollution [6]. For instance, this type of wastewater is a significant source of phosphorus in surface water bodies. Since it can be the limiting nutrient for plant growth, phosphorus removal is important in order to prevent eutrophication [15,16]. Even though loads from the point sources to surface water bodies have been reduced due to improvement in wastewater treatment, non-point source phosphorus pollution has increased because of intensified agricultural production [15]. Apart from phosphorus, another important surface water pollutant originating from agricultural non-point sources is nitrate, whose level has to be safely controlled and reduced according to the EU directive 91/676/EEC [17].

Since wetlands can reduce agricultural pollution [18], a broader implementation of CWs has been proposed as a measure for protecting aquatic systems [16,19]. The CW type that is usually applied for agricultural drainage water treatment is the surface flow constructed wetland (SFCW) [4], which has been reported to be able to remove nitrogen and phosphorus at relatively low cost [14]. Although there have been studies that discuss different aspects of this topic, not many of them present the general functioning of these systems over long periods of time.

Therefore, the aim of this study, which was carried out in northern Italy, was to review the results of long-term monitoring of a SFCW that had been in operation under a variety of conditions in terms of its ability to treat agricultural drainage water and the effect of such a long operational period on the CW soil and plant characteristics. The hypothesis of this research was that SFCWs can be considered as a long-term viable option for agricultural drainage water treatment.

2. Materials and Methods

The study was carried out on a non-waterproofed SFCW located on an experimental agricultural farm (44°34′22.2″ N, 11°31′44.9″ E) of the Canale Emiliano Romagnolo (CER) Land Reclamation Consortium near Budrio village in Emilia-Romagna region (Italy; Figure 1). According to the Köppen climate classification, the climate of the site is subhumid, with a mean annual rainfall of 771 mm and mean annual temperature of around 13.7 °C. The site receives most of its precipitation during spring and autumn. The coldest month is January (mean temperature 2.7 °C) and the warmest is July (mean temperature 24.3 °C). These precipitation and temperature data are 30-year normal values supplied by ARPAE (Regional Agency for Environmental Protection) from the nearest weather station [20]. However, the climate data used for the calculations in this paper were taken from a weather station managed by CER and located about 500 m from the SFCW.

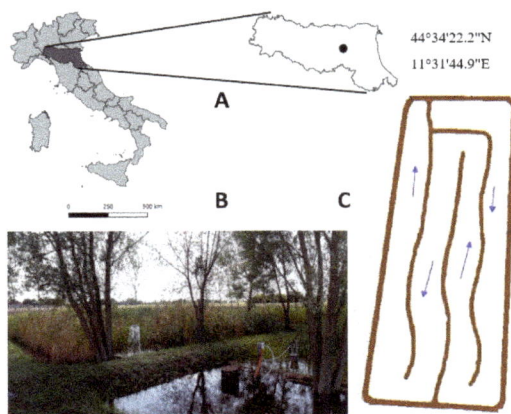

Figure 1. The location of the SFCW system in Emilia-Romagna region (Italy) (**A**), constructed wetland studied (**B**) and its hydraulic scheme (arrows show the direction of the water flow) (**C**).

The CW treats tile drainage water coming from a 12.5 ha experimental farm that grows different crops (e.g., fruit trees, vegetables, and cereals) throughout the year. The area of the SFCW represents around 3% of the total farm surface, and it is divided into four 8–10 m wide meanders that create a 470-m-long water course (Figure 1). The total capacity of the wetland is about 1500 m^3 when the water level reaches 0.4 m, which is the maximum water height below the discharge level. The whole farm area is drained to a main ditch from which water is conveyed into the CW by means of two pumps. The CW effluent is discharged into a canal, which collects excess water from the fields of other neighbouring farms.

2.1. Experimental Design

Since its construction in 2000, the CW has functioned continuously; however, its monitoring was done from 2003 to 2009, because of different projects conducted at the farm. In 2017, the monitoring started again, thanks to an Italian National Research Project (Green4Water) coordinated by the University of Bologna.

The operation of the SFCW depends mostly on the frequency and volume of precipitation. However, in the past its operation has sometimes been modified, and the system exposed to different feeding conditions, because of specific research requirements, as indicated in the following. During summer 2004, an unusually high inflow (around 3300 m^3) entered the system, since a specific experiment regarding nutrient removal was being conducted. Additionally, in 2007 and partly in 2008, the influent and effluent pumps were often turned off in order to allow plantation of willow and poplar as the farm participated in a project related to biomass production.

2.1.1. Water Balance and Quality

The system is equipped with two mechanical flow meters that record influent and effluent volumes every hour, and two automatic samplers that take influent and effluent water samples on the basis of flow from the beginning of the event (at specific inlet water volume values) and time (every 24 h), respectively. Since the functioning of the system depends mostly on the presence of precipitation, no general sampling schedule was established, and sometimes no samples were taken for a few weeks due to the lack of drainage water. The water level inside the CW is measured by a specific sensor. All the collected data are managed and recorded by a central control system. The precipitation height data are taken from the farm weather station, which is also equipped with a precipitation sampling unit.

Since August is generally the month with the lowest inflow volume, the hydrological year was defined as the period between 1st September and 31st August. TN and TP, as well as NO_3^- and NH_4^+ balances, were defined for six hydrological years (from 2003–2004 to 2008–2009). The SFCW dynamic water budget [1] can be expressed as:

$$Q_{in} + (P \times A) - Q_{out} - I - (ET \times A) = \frac{dV}{dt} \qquad (1)$$

where:

Q_{in} = inflow rate (m$^3 \cdot$ d^{-1});
P = precipitation rate (m·d^{-1});
A = wetland top surface area (m^2);
Q_{out} = outflow rate (m$^3 \cdot$ d^{-1});
I = infiltration flow rate (m$^3 \cdot$ d^{-1});
ET = evapotranspiration rate (m·d^{-1});
V = water storage inside the SFCW (m^3);
t = time (d).

Over long averaging periods (Δt), the change in storage (ΔV) can be considered negligible [1]. Moreover, it was not possible to measure infiltration and evapotranspiration rates separately. Therefore, a simplified water balance over each hydrological year was calculated as:

$$Q_{in} + (P \times A) - Q_{out} = I + (ET \times A) \tag{2}$$

In Equation (2) the term $(I + (ET \times A))$ was considered as the overall water loss from the SFCW. Accordingly, the outflow/inflow ratio (R, %) was expressed as:

$$R = \frac{Q_{out}}{(P \times A) + Q_{in}} \times 100 \tag{3}$$

Equation (3) represents the percentage of influent water flowing out of the system through the outlet device, while the complementary percentage is assumed to be the amount of water jointly lost due to infiltration and evapotranspiration.

Water samples (i.e., precipitation and CW influent and effluent) were analysed for total nitrogen (TN), nitrate (NO_3^-), ammonium (NH_4^+), and total phosphorus (TP). TN was measured by the elemental analyser Shimadzu TNM-1 (Shimadzu, Kioto, Japan), and NO_3^- and NH_4^+ concentrations were determined by using a flow analyser (AA3, Bran Luebbe, Norderstedt, Germany). TP analysis was performed by using an inductively coupled plasma optical emission spectrometer ICP-OES which was equipped with a plasma source and an optical detector with a charge-coupled device CCD (SPECTRO Analytical Instruments GmbH & Co., Kleve, Germany). Water samples were analysed for TN after addition of 1% HNO_3 (>69% *v/v*, for trace analysis, Sigma-Aldrich, St. Louis, MO, USA). Before analysis, all samples were filtrated through Watman 42 filters (Merck KGaA, Darmstadt, Germany).

The samples taken and analysed for nutrient concentrations at inlet (C_{in}, mg·L^{-1}) and outlet (C_{out}, mg·L^{-1}) were multiplied by the related measured volume of water that flowed into (V_{in}, m^3) or out (V_{out}, m^3) of the system, respectively. Afterwards, all the inflow and outflow loads during one hydrological year were summed to calculate the mass of nutrients (kg·year^{-1}) entering and exiting the wetland yearly. On this basis, yearly mass retention rate (RR, %), over each hydrological year, was calculated as:

$$RR = \frac{\Sigma V_{in} \times C_{in} - \Sigma V_{out} \times C_{out}}{\Sigma V_{in} \times C_{in}} \times 100 \tag{4}$$

2.1.2. Vegetation

The SFCW has been continuously inhabited by different aquatic plant species, such as common reed (*Phragmites australis*) or cattail (*Typha latifolia*, *T. angustifolia*). In autumn 2006, after the removal of the above-ground biomass of the aquatic plants from the first two meanders, willow (*Salix alba*) and poplar (*Populus alba*) were planted, as requested by the project on biomass production. With the exception of this partial removal, the plant biomass has never been harvested.

In the 2004–2009 period, vegetation was sampled once a year, approximately at the middle of each of the four meanders, for nutrient and heavy metal content evaluation. The samplings were performed at the end of October/beginning of November, so those results roughly correspond to the previous hydrological year, which ended in August, 2–3 months before the plant sampling.

Biomass dry weight, average height and number of shoots per m^2 were measured. Additionally, biomass samples were divided into below- and above-ground biomass, and were tested for TN and TP content, as well as for the presence of the semi-metal boron (B) and different heavy metals (cadmium (Cd), chrome (Cr), copper (Cu), iron (Fe), lead (Pb), manganese (Mn), nickel (Ni) and zinc (Zn)).

TN analysis of vegetation was performed by using a thermo-electron CHNS-O elemental analyser (Thermo Fisher Scientific, Waltham, MA, USA). TP and metals were measured using ICP-OES. Before elemental analysis, biomass samples were dissolved in a mixture of HNO_3 (>69% *v/v*, for trace analysis, Fluka, Sigma-Aldrich, St. Louis, MO, USA) and H_2O_2 (30% *v/v*, for trace analysis, VWR Prolabo Chemicals, Radnor, PA, USA) in a ratio of 4:1 (*v:v*) by microwave-assisted digestion (Start D, Micro-wave Digestion System, Milestone srl, Bergamo, Italy).

2.1.3. Soil

The first 15 cm of soil was sampled once a year (at the end of October/beginning of November) in the period 2004–2009, approximately at the middle of each of the four meanders. In July 2017, core soil samples (Figure 2) were taken up to 60 cm depth in order to see how the 17 years of CW operation had affected the soil composition along its profile. After manual removal of plant roots up to a diameter of ca. 1–2 mm, the samples were air-dried and sieved to 2 mm. Afterwards, they were tested for organic matter, TN and TP content, as well as for metal content. No data on metals and nutrients in the CW soil before 2004 is available as a control.

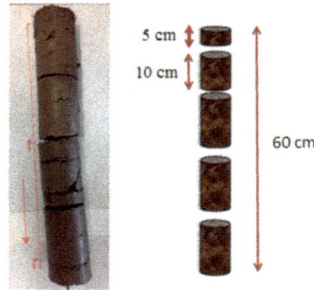

Figure 2. The soil core sampling and sectioning.

Total organic carbon (TOC) and TN in the soil samples were determined by using a thermo-electron CHNS-O elemental analyser (Thermo Fisher Scientific, Waltham, MA, USA). TP and metal concentrations were analysed by using ICP-OES, after dissolution of soil samples in a mixture of HCl (37% v/v for trace analysis, Sigma-Aldrich, St. Louis, MO, USA), HNO$_3$ (>69% v/v, for trace analysis, Fluka, Sigma-Aldrich, St. Louis, MO, USA), and H$_2$O$_2$ (30% v/v, for trace analysis, VWR Prolabo Chemicals, Radnor, PA, USA) in the ratio of 4:1:0.25 (v:v:v) by microwave-assisted digestion (Start D, Micro-wave Digestion System, Milestone, MD, USA).

2.2. Data Analysis

The correlation of inflow water volume and precipitation was checked with the Pearson Product Moment Correlation test. The water influent and effluent concentrations of different parameters considered were compared with Student's T-test, while the content of different compounds at different depths along the soil profile in 2017 were compared with ANOVA. Before those tests were performed, the data was checked for normality and equal variance, and if the assumptions were not met, the values were log$_{10}$ transformed. In cases where the assumptions were not met even after the transformation, the Mann-Whitney U test and ANOVA on ranks were used. All analyses were performed using SigmaPlot 14 software (Systat Software, Inc, San Jose, CA, USA).

3. Results and Discussion

3.1. Simplified Water Balance

The water balance (cumulated yearly inflow and outflow volumes) and the outflow/inflow ratio (calculated as in Equation (3)) of the monitored CW are shown in Figure 3. With the exclusion of the 2006–2007 and 2007–2008 hydrological years, when the pumps were often turned off, the inflow was in the range 8400 to 19,500 m^3·year^{-1}, while the outflow was between 750 and 9500 m^3·year^{-1}. Such a high yearly variation between inflow and outflow of the system was due to precipitation differences during the years, as well as the specific functioning conditions required by the different research projects, as already explained in Section 2.1.1. On the other hand, the irregular water flow in the system and water loss can be better seen through the reported outflow/inflow ratio, which ranged between 9 and 65% (Figure 3). Since the SFCW was not waterproofed, such a high loss could mainly be attributed to the infiltration rate. However, it was also related to high evapotranspiration values, especially during the summer months, as already reported by Borin et al. for Northern Italy [21].

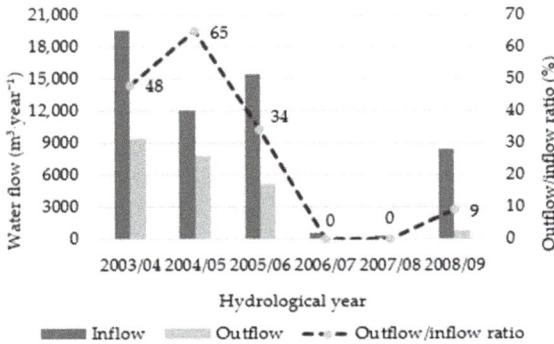

Figure 3. Water balance of the monitored SFCW.

The two SFCWs comparable to the system discussed in this study were located in the USA and had total volumes of 5400 and 1200 m^3 [19]. The outflow/inflow ratios of these systems were in the range 0–60%, indicating considerable water losses that were mostly attributed to infiltration. However, the operational variability of SFCW systems is very wide, and they were reported to function with higher outflow/inflow ratios [12,18] and different inflow rates [15,16]. Therefore, SFCWs can be considered to be flexible and adaptable to different hydrological conditions and loading rates. These characteristics will be discussed in the context of nutrient retention in the next section.

Of the six hydrological years considered, 2005–2006 was selected as representative of annual water flow in the system. As can be seen in Figure 4, the majority of inflow into the system took place during autumn (October and November) and early winter (December), when the greatest part of yearly precipitation also occurred. Although the precipitation that fell onto the farm area took a certain time to infiltrate and reach the SFCW, it was still significantly correlated with the inflow (r = 0.661), as shown with the Pearson Product Moment Correlation. The missing or very low outflow from the system in the period January–August 2006 is related to the low inflow. Since the maximum capacity of the SFCW studied was about 1500 m^3, inflows lower than the same value cannot usually cause considerable outflow (depending on the water level inside the SFCW). The water stored inside the system was later partly lost to infiltration and evapotranspiration.

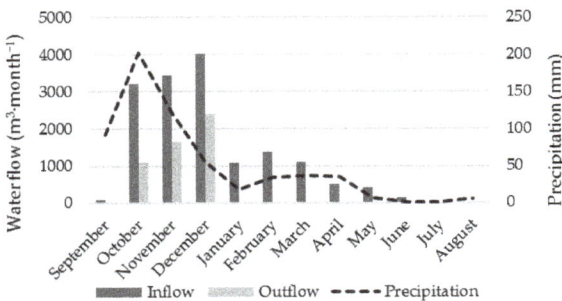

Figure 4. SFCW water flow and precipitation in the hydrological year 2005–2006.

3.2. Water Quality

Fertilisation was the main source of nitrogen that entered the farm (see Table S1 in supporting information). Since most of the applied TN (>80%) was not present in the drainage water, it can be inferred that in the field, the nutrient was uptaken by crops and retained by the soil components, although its release to the atmosphere in volatile forms cannot be ruled out.

Table 1 gives average influent and effluent concentrations of the measured parameters during the hydrological years considered. The TN influent concentration was in the range 13.7–21.9 mg·L^{-1}, if the hydrological years 2006–2007 and 2007–2008 (when the pumps were mostly turned off) are excluded. If the same two years are not considered, nitrate influent concentrations did not vary a lot. However, nitrate effluent concentration in 2005–2006 was considerably lower than in 2004–2005. On the other hand, both influent and effluent concentrations of ammonium and TP were low (<0.4 mg·L^{-1}). The significant differences ($p < 0.05$) between influent and effluent concentration were found only for TN and nitrate in the hydrological year 2005–2006 (Table 1). This might be connected to the high water loss during that year (Figure 3), which could be related to high infiltration and, therefore, nutrient loss. However, another factor that should be considered is increased residence time in the system, and thus the increased removal of pollutants.

Table 1. Average concentrations of different parameters (mg·L^{-1}) at the inlet and outlet of the monitored system (values are displayed as: mean ± st. error and number of samples in brackets).

Hydrological Year	Parameter	Total Nitrogen	Nitrate	Ammonium	Total Phosphorus
2003–2004	Inflow	21.9 ± 2.7 (34)	-	-	-
	Outflow	14.5 ± 2.4 (17)	-	-	-
	T test-p value	0.078	-	-	-
2004–2005	Inflow	16.5 ± 1.4 (27)	14.2 ± 1.4 (27)	0.1 ± 0.0 (27)	0.1 ± 0.0 (27)
	Outflow	13.4 ± 1.3 (33)	11.7 ± 1.5 (33)	0.3 ± 0.1 (33)	0.4 ± 0.1 (33)
	T test-p value	0.124	0.229	0.272	0.083
2005–2006	Inflow	13.7 ± 1.2 (17)	11.4 ± 1.1 (17)	0.1 ± 0.0 (17)	0.1 ± 0.0 (17)
	Outflow	8.2 ± 1.2 (11)	4.5 ± 0.8 (11)	0.1 ± 0.0 (11)	0.0 ± 0.0 (11)
	T test-p value	**0.004**	**<0.001**	0.226	0.353
2006–2007	Inflow	2.4 ± 0.6 (2)	0.8 ± 0.1 (2)	0.1 ± 0.0 (2)	0.2 ± 0.2 (2)
	Outflow	*	*	*	*
	T test-p value	-	-	-	-
2007–2008	Inflow	8.7 (1)	-	-	0.2 (1)
	Outflow	*	*	*	*
	T test-p value	-	-	-	-
2008–2009	Inflow	17.6 ± 2.1 (11)	-	-	-
	Outflow	11.2 ± 2.7 (10)	-	-	-
	T test-p value	0.071	-	-	-

Notes: T test p values show the statistical comparison (by T-test or Mann-Whitney U test) of the influent and effluents in the given year. Bolded values show significant difference. - Undetected. * No outflow.

As shown in Tables 1 and 2, nitrate was the most prevalent nitrogen species in the TN load in the influent to our CW. The retention rate (*RR*, %) of the TN load by the system varied during the six hydrological years monitored, but it never dropped below 47%. Comparing the data reported in Table 2 and Figure 3, it can be seen that the TN retention depended on the outflow/inflow ratio, rather than on the mass load of nitrogen that entered the CW. These findings are in accordance with Groh et al. [19], who concluded that nitrate removal mostly depends on the hydraulic loading, and who reported similar removal of TN. Retention rate of TP in the hydrological year 2005–2006 was 49%, which is higher than the 36% retention reported for a Swedish SFCW treating agricultural drainage water [12]. However, since the SFCW studied was not waterproofed, it is expected that some of the nutrients infiltrated through the ground, and therefore were not removed, but rather retained.

The lowest retention rate of TN, nitrate, ammonium, TP parameters in 2004–2005 can be explained by the highest outflow/inflow ratio, which consequently means shorter residence time. Similarly, in December 2005 (the month with the highest TN and NO_3^- load; data not shown, but included in Table 2: 2005–2006 nutrient balance), inflow and outflow volumes were comparable. thus suggesting a very short residence time. In addition, a similar hydrological situation repeated in March–April 2006 and during both months, TN and TP outflow loads were higher than the inflow ones (data not shown, but included in Table 2: 2005–2006 nutrient balance), thus suggesting that sediments containing these nutrients were partially flushed out.

Table 2. Nutrient balance of the monitored CW with the atmosphere input considered.

Balance	Parameter	2003–2004	2004–2005	2005–2006	2006–2007	2007–2008	2008–2009
TN	Input (kg·year^{-1})	313	216	219	6	12	180
	Output (kg·year^{-1})	125	115	39	0	0	11
	Retention rate (%)	60	47	82	100	100	94
NO$_3^-$	Input (kg·year^{-1})	-	182	180	0.5	-	-
	Output (kg·year^{-1})	-	99	21	*	*	-
	Retention rate (%)	-	45	88	100	-	-
NH$_4^+$	Input (kg·year^{-1})	-	1.1	1.3	0.1	-	-
	Output (kg·year^{-1})	-	1.1	0.4	*	*	-
	Retention rate (%)	-	0	71	100	-	-
TP	Input (kg·year^{-1})	-	1.2	1.4	0.1	0.1	-
	Output (kg·year^{-1})	-	2.1	0.7	0.0	0.0	-
	Retention rate (%)	-	-	49	100	100	-

Notes: - Undetected. *No outflow.

The formation of phosphorus species bound to the sediment components is one of the main mechanisms of TP removal from water in SFCWs [12]. It is likely that the rather high outflow/inflow ratio in 2004–2005 (Figure 3) mobilised the finest sediment components, thus explaining why the system acted as a TP source rather than a sink (Table 2). The hypothesised phosphorous dynamics can be confirmed by the observations reported in the following section.

3.2.1. Analysis of Single Events

Since six hydrological years is a long period of time, it can be useful to look at single events. Two of them (both 25 days long) were selected in order to stress the SFCW's efficiency in nutrient retention: the first one from 10th November to 5th December 2004 and the second one from 6th to 30th November 2005. As shown in Table 3, the system generally performed better during the latter one, despite the higher water flow.

Table 3. Water and nutrient balance during two different events.

Parameter	10th November–5th December 2004			6th–30th November 2005		
	Inflow	Outflow	Retention	Inflow	Outflow	Retention
Water volume (m^3)	1138	1086	-	2857	1223	-
TN (kg)	15.17	4.69	59%	41.63	10.20	75%
NO$_3^-$ (kg)	11.52	1.64	85%	36.57	6.55	82%
NH$_4^+$ (kg)	0.06	0.57	-	0.13	0.07	45%
TP (kg)	0.03	0.57	-	0.12	0.07	43%

Note: - Undetected.

Water inflow and outflow during the first event were similar (1138 and 1086 m^3, respectively). On the contrary, during the second event the outflow (1223 m^3) was considerably lower than the inflow

(2857 m^3) (Table 3). This behaviour can be explained by the fact that, before the first event, the water level in the system was 35.7 cm, whereas before the second one, it was only 5 cm. Since the water level was close to its maximum value of 40 cm, the water already contained in the system at the beginning of the first event shortened the water retention time, thus causing lower compound retention rates. The sudden flush-out of the soil particles can explain the higher TP outflow than the inflow load [12].

3.3. CW Soil and Vegetation

3.3.1. Plant Development

The agronomic results of plant sampling are shown in Table 4. As already stated, plant species in the SFCW changed over the course of time due to the different conditions requested by some projects that the CER Land Reclamation Consortium participated in. The contained variation of plant biomass observed in the 2004–2006 period, ranging from 4.68 to 5.40 kg·dw·m^{-2}, has to be considered in the context of the equilibria among the different aquatic plant species inhabiting the system. In this period, TN and TP that entered the CW through agricultural drainage water enabled plant development. The observed above-ground biomass is in line with that reported by other studies, for example, [22] recorded 0.37–1.76 kg·dw·m^{-2} and [23] 1.2–1.4 kg·dw·m^{-2}.

Table 4. Agronomic results of plant sampling.

Plant Survey Parameter	2004	2005	2006	2007	2008	2009
Main species	P, T	P, T	P, T	P, T, W, L	P, T, W, L	P, T, W, L
Dry weight (kg·m^{-2})	5.32	5.40	4.68	3.02	7.03	5.63
above-ground	3.01	2.74	1.57	0.21	0.70	0.88
below-ground	2.31	2.66	3.11	2.81	6.33	4.75
above/below ground ratio	1.30	1.03	0.50	0.07	0.11	0.19
Average height (cm)	247	253	190	131	239	301
Shoots (number m^{-2})	223	171	210	73	105	57

Notes: P—*Phragmites*; T—*Typha*; W—Willow; L—Poplar.

Over the years, there was a decrease in the ratio between above and below-ground biomass. While in 2004 it was 1.3, in 2009 it dropped to 0.2, meaning that below-ground biomass had 5 times greater weight than above-ground biomass. This observation is likely due to the fact that above-ground biomass was never harvested, so withered plants from the previous year blocked the sunlight for sprouting plants, took space and consequently limited development of new shoots [23].

The visible drop of all biomass parameters, including biomass, average height and number of shoots, in 2007 (Table 4) is mainly due to the removal of above-ground biomass from the first two meanders, as clearly indicated by the low value of the above-ground/below-ground biomass ratio, as well as the interruption of water inflow and nutrient input into the system to allow for the poplar and willow plantation/rooting. After 2007, the agronomic parameters indicated the establishment of new plant consortia.

3.3.2. Nutrients' Content of Biomass and Soil

Owing to the removal of above-ground biomass of aquatic plants from the first two meanders in autumn 2006, no nutrient distribution can be estimated between soil and vegetation. Nevertheless, some considerations on their occurrence are still possible.

The plant biomass sampled was analysed for its TN and TP content, as shown in Figure 5. In the 2004–2009 period, the content of both of these elements in the biomass show some fluctuations that could be related to the variation in inflow water (Figure 3) and nutrient loads (Table 2), as well as to the change in plant species that inhabited the SFCW. The plantation of willow and poplar trees in 2006 surely increased plant competition, thus modifying the equilibria among plant and microorganism species.

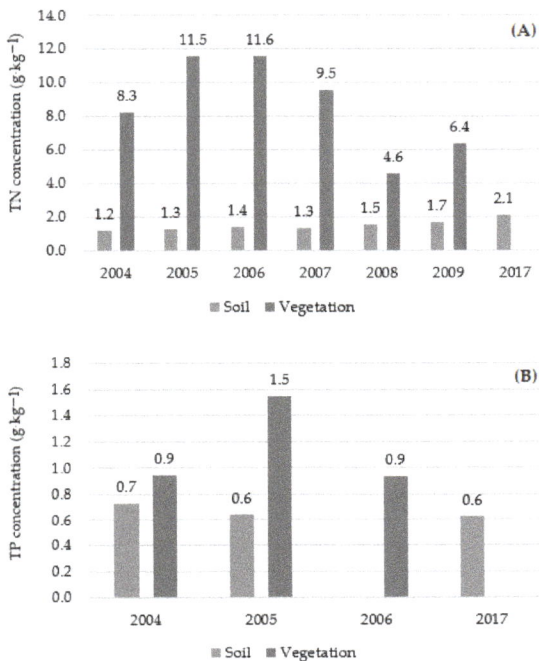

Figure 5. TN (A) and TP (B) content of the surface soil layer and vegetation in the SFCW.

Only partial TP data were available for the same period. Nevertheless, the content of plant biomass peaked in 2005 at 1.5 g·kg^{-1} (Figure 5). Similarly, the biomass TN content peaked in the 2005–2006 period at about 11.5 g·kg^{-1} (Figure 5). This value is a bit lower than the TN reported by [22] in a study where a SFCW received municipal wastewater that was more polluted than the agricultural drainage water used in this study.

TN and TP concentrations in the SFCW top soil are reported in Figure 5. Although no values for nutrients or for OMC (organic matter content) in CW soil are available from before 2004 to be considered as a reference, it is possible to observe that TN concentration slowly increased over the years, reaching a value of 2.1 g·kg^{-1} in 2017, almost twofold higher than that of 1.2 g·kg^{-1} measured in 2004. TP, on the other hand, did not accumulate in the top soil, as is evident by its quite constant level over the 13-year-long observation period. This can be explained by the low nutrient loads of the influent (Table 2) and the specific flush-out events already described in the Section 3.2.1.

As far as the OMC of the SFCW soil is concerned, an increase in the first 15 cm was observed since the beginning of the monitoring. While the OMC in 2004 was 19.6 g·kg^{-1}, it increased more than 2.5 times in 2017, reaching a value of 49.8 g·kg^{-1}. This positive trend suggests constant organic matter production and its sedimentation into the CW during the 14 years of the system's operation. The higher OMC increase in other SFCW soils reported in the literature is typical of applications of water containing higher amounts of organic matter and nutrients than those contained in our agricultural drainage water. For example, [24] reported a tenfold increase of TOC over a period of 5 years, but in this case several applications of slurry were made to the SFCW.

3.3.3. Boron and Heavy Metals in Biomass and Soil

Additional information on the state of the CW can be obtained by considering the metal content of the biomass and soil, as these elements, apart from being naturally contained in the soil in background levels, can enter the SFCW by water inflow and can be subsequently uptaken by plants or accumulated in soil. Cu and Zn are present in several plant products, either as active ingredients themselves or as counterions of organic products [25], so their inflow to the SFCW can be seasonal or constant, depending on their administration frequency. B is a plant micronutrient usually applied to crops as fertiliser [26]. Cd, Ni and Pb are considered potentially toxic elements (PTE) and can occur in wastewater and be accumulated in soil through different anthropogenic activities [27].

Figure 6 gives the content of these metals in the above- and below-ground biomass for the period 2004–2006, before the plantation of new trees in the SFCW. As a general trend, it is possible to observe that the metals were found to mostly have accumulated in the below-ground plant tissues, and only a small portion was transferred to the above-ground parts. These findings are in accordance with [22], which also reported accumulation of heavy metals in below-ground biomass.

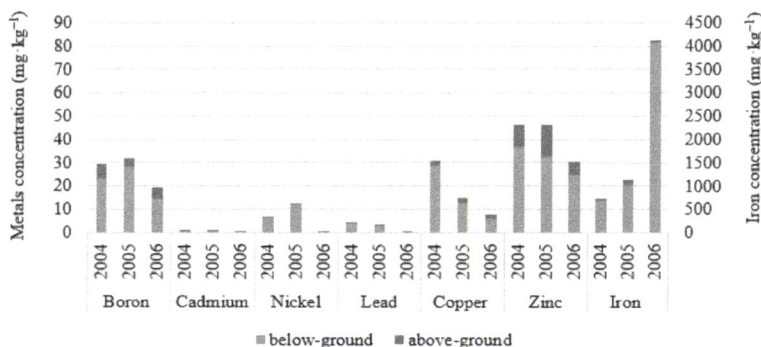

Figure 6. Heavy metal concentrations in below- and above-ground plant biomass.

Furthermore, high variation was observed among the amounts of heavy metals retained by plant tissues. The metal accumulated the most by plants was Fe, with a maximum concentration of more than 4000 mg·kg^{-1} in 2006. Cd was never higher than 0.8 mg·kg^{-1}, Cu ranged between 31 and 8 mg·kg^{-1}, while Zn was never less than 30 mg·kg^{-1} (Figure 6). Similar results were reported by [28] for *Phragmites australis* from different horizontal flow CWs. For example, slightly higher Zn levels and a few times higher Cu concentration in *Typha* were reported after a 15-day-long experiment [13]. As far as B is concerned, this nutrient was found to be well retained by plant root apparatus (27 mg·kg^{-1} on average) in the 2004–2006 period.

Some considerations can be drawn by comparing the concentrations of heavy metals in the biomass with their levels in the top soil (Table 5) for the period 2005–2006, when a complete set of data is available. In fact, even though the root system of aquatic plants can enter the soil deeper than 15 cm, the level of heavy metals in the top soil gives some indications as to their accumulation in the system and their potential bioavailability to plants.

Table 5. Concentration of heavy metals (mg·kg^{-1}) in below-ground biomass and the top soil layer.

Parameter	Biomass			Soil		
	2005	2006	Average	2005	2006	Average
Cadmium	0.4	0.0	0.2	0.0	0.0	0.0
Copper	12.7	6.1	9.4	35.5	32.0	33.8
Iron	1014.7	4086.2	2550.5	29.5	40.9	35.2
Lead	3.0	0.3	1.7	37.0	13.5	25.3
Nickel	12.2	0.2	6.2	54.5	54.0	54.3
Zinc	32.5	24.9	28.7	109.5	85.0	97.3

In general, the concentrations of heavy metals in the first 15 cm of the CW soil were on average lower than, but still in the range of, other studies, such as, for example, that reported by [29], who studied heavy metal removal using a horizontal flow CW from road runoff in Ireland.

In detail, in the 2005–2006 observation period, Cd was present at very low concentrations in both soil and biomass, whereas the average Pb level in the top soil of 25.3 mg·kg^{-1} and its relatively low concentration in the biomass (1.7 mg·kg^{-1}, Table 5) are in accordance with the fact that it is a fairly immobile element [27], as well as its known low bioavailability to plants [30]. The other heavy metals, including Ni, Cu and Zn, are considered of medium bioavailability to plants in aerated soil [30]. In the system reported in this study, the average biomass concentrations of Ni, Cu, and Zn (6.2, 9.4 and 28.7 mg·kg^{-1}, respectively) were lower than, but still in the range of, their levels in the top soil (54.3, 33.8 and 97.3 mg·kg^{-1}, respectively), thus indicating moderate availability to plants.

The Fe level in the biomass (2550.5 mg·kg^{-1}) was very high, and the top soil concentration of 35.2 mg·kg^{-1} cannot justify such a high accumulation at the roots. Specific root uptake mechanisms have to be considered for this metal, as already defined by [31], who have reported that aquatic plants may modify the rhizosphere by facilitating the formation of iron oxide plaques that immobilise and concentrate heavy metals.

Under conventional operational conditions, the aquatic plants of the system studied act as Fe accumulators and Cu, Pb, Ni and Zn bio-indicators. These findings could be of interest for further study on metal bioavailability for aquatic plants and phytoremediation mechanisms against PTE.

Finally, the actual SFCW state was monitored as well, in order to evaluate how the 17 years of operation had affected the distribution of nutrients and heavy metals along the soil profile (Table 6). As expected, the level of total organic carbon, TN, TP and B decreased from the top soil to the deeper layers, even though significant differences ($p < 0.05$) were shown only for the first two parameters (Table 6). This highlights the surface accumulation of nutrients and organic material produced by aquatic plants. The average pH value of the soil (measured in water) was found to be 8.51 ± 0.04 (20).

No visible accumulation of heavy metals can be seen along the vertical soil profile, and there were no significant differences among them (Table 6). Most probably, this is due to the fact that the root systems of aquatic plants can accumulate large amounts of heavy metals [32] and occupy large soil volumes. Moreover, when compared to their legal limits, as introduced by the Italian law [33], all of the heavy metals resulted at concentrations below the lowest admitted threshold (limit A: soil suitable for private and public green areas). In addition, the data for Cr, Ni, Zn, Cu, Pb, and Sn of CW soil still safely fall within the range of the local anthropic-natural available background data ($\leq$75, $\leq$120, $\leq$75; $\leq$60, and $\leq$50 mg·kg^{-1}, respectively [34]), thus indicating that no important changes have affected the soil heavy metal content. Therefore, even after 17 years of functioning, the SFCW can still be considered to be a bio-filter with no visible accumulation of PTE in either the top or the deep soil layers (up to 60 cm depth), and with levels below the current legal limits.

Table 6. Concentrations of nutrients or heavy metals at different depths of the CW soil in 2017 (values are displayed as: mean ± std error and number of samples in brackets).

Parameter	0–5 cm	5–15 cm	15–30 cm	30–45 cm	45–60 cm	Limit A	Limit B
Total organic carbon (g·kg⁻¹) *	41.5 ± 6.1 (4) [a]	16.3 ± 1.7 (4) [b]	10.2 ± 0.5 (4) [b,c]	8.4 ± 1.1 (4) [c]	7.2 ± 1.3 (4) [c]	-	-
Total nitrogen (g·kg⁻¹) *	3.5 ± 0.4 (4) [a]	1.6 ± 0.1 (4) [b]	1.2 ± 0.0 (4) [b,c]	1.0 ± 0.1 (4) [b,c]	0.9 ± 0.2 (4) [c]	-	-
Total phosphorus (g·kg⁻¹)	0.7 ± 0.0 (4)	0.6 ± 0.0 (4)	0.6 ± 0.1 (4)	0.5 ± 0.0 (4)	0.5 ± 0.1 (4)	-	-
Boron (mg·kg⁻¹)	44.5 ± 0.6 (4)	45.3 ± 0.3 (4)	44.0 ± 0.4 (4)	42.3 ± 1.7 (4)	40.5 ± 2.1 (4)	-	-
Cadmium (mg·kg⁻¹)	0.2 ± 0.0 (4)	0.2 ± 0.0 (4)	0.1 ± 0.0 (4)	0.2 ± 0.0 (4)	0.1 ± 0.0 (4)	2	15
Chrome (mg·kg⁻¹)	67.9 ± 0.8 (4)	72.0 ± 0.6 (4)	71.6 ± 1.0 (4)	68.4 ± 1.5 (4)	66.4 ± 2.1 (4)	150	500
Copper (mg·kg⁻¹)	36.9 ± 1.6 (4)	36.6 ± 1.4 (4)	38.9 ± 4.5 (4)	35.0 ± 3.5 (4)	32.2 ± 4.0 (4)	120	600
Iron (g·kg⁻¹)	24.9 ± 0.5 (4)	26.5 ± 0.3 (4)	26.5 ± 0.1 (4)	25.9 ± 0.4 (4)	25.0 ± 0.7 (4)	-	-
Lead (mg·kg⁻¹)	24.8 ± 2.6 (4)	23.8 ± 1.2 (4)	24.9 ± 2.0 (4)	22.9 ± 1.4 (4)	20.0 ± 1.1 (4)	100	1000
Nickel (mg·kg⁻¹)	48.5 ± 0.7 (4)	51.2 ± 0.6 (4)	51.2 ± 0.2 (4)	49.5 ± 1.1 (4)	48.3 ± 1.3 (4)	120	500
Zinc (mg·kg⁻¹)	78.3 ± 1.4 (4)	77.1 ± 1.6 (4)	77.0 ± 2.6 (4)	73.7 ± 2.7 (4)	70.6 ± 2.8 (4)	150	1500

Notes: Limit A—green areas, private and residential use; Limit B—commercial and industrial use (Italian D.Lgs. 152, 2006). * ANOVA test showed significant differences only for Total organic carbon and Total nitrogen. Letters for concentrations of these two parameters at different depths indicate significant ($p < 0.05$) difference (different letter) or not (the same letter). - Undetected.

4. Conclusions

This study presents the main findings of monitoring performed over a long period of time (2003–2017) in the operation of a SFCW located at the farm of the CER Land Reclamation Consortium (Northern Italy), treating agricultural drainage water. Its functioning was particular, and it depended on the needs of the farm and specific research projects, and was therefore receiving different water volumes with different nutrient/pollutant loads throughout the years.

The retention of TN and TP, the two nutrients that are mostly responsible for the surface water bodies' pollution and eutrophication, was never below 47% and 49%, respectively. Since the SFCW received varied inflow loads over the years, it can be said that the system proved itself to be a viable option for tile drainage water treatment. As a general rule, TN and TP retention depended on their residence time in the system, rather than their inflow loads. Even though the loads of these nutrients varied a lot over the years, their analysis in the biomass and soil showed a certain accumulation.

In addition, heavy metals that entered the SFCW were mostly retained by the root system, thus acting as a biofilter for the collected agricultural drainage water and protecting the receiving water bodies. For several heavy metals, the distribution between biomass and soil made it possible to define their varying bioavailability to the aquatic plants inhabiting the system and showed their potential for the removal of these PTEs. Finally, and most importantly, after 17 years of functioning, the SFCW soil content of each of the heavy metals considered was found to be below the lowest limit imposed by the Italian law for soils of private and public green areas.

In light of these observations, it is possible to conclude that the monitored SFCW was able to adapt its performance and ecosystem services to different operational conditions over a long period of time, without losing its ability to improve the inflow water quality.

Supplementary Materials: The following are available online at http://www.mdpi.com/2073-4441/10/5/644/s1, Table S1: Farm TN input and SFCW nitrogen balance (all the values are in kg).

Author Contributions: Conceptualization, A.T., S.L. and I.B.; Methodology, A.T., S.L., S.B., S.A. and P.M.; Laboratory Analysis, I.B., S.B. and D.S.; Field Investigation, S.A. and D.S.; Data Curation, S.L., S.B., S.A. and D.S.; Writing-Original Draft Preparation, S.L. and I.B.; Writing-Review & Editing, S.L., I.B. and A.T.; Supervision, A.T.; Funding Acquisition, A.T. and P.M.

Funding: This research was funded by Regione Emilia-Romagna, Autorità di Bacino del Reno and by Ministero dell'Istruzione, dell'Università e della Ricerca (MIUR) through the Research Project Green4Water (grant number: PRIN2015AKR4HX) available at https://site.unibo.it/green4water.

Conflicts of Interest: The authors declare no conflict of interest.

References and Notes

1. Kadlec, R.H.; Wallace, S.D. *Treatment Wetlands*, 2nd ed.; CRC Press: Boca Raton, FL, USA, 2009.
2. Meng, P.; Pei, H.; Hu, W.; Shao, Y.; Li, Z. How to increase microbial degradation in constructed wetlands: Influencing factors and improvement measures. *Bioresour. Technol.* **2014**, *157*, 316–326. [CrossRef] [PubMed]
3. Carvalho, P.N.; Arias, C.A.; Brix, H. Constructed Wetlands for Water Treatment: New Developments. *Water* **2017**, *9*, 397. [CrossRef]
4. Leon, A.S.; Tang, Y.; Chen, D.; Yolcu, A.; Glennie, C.; Pennings, S.C. Dynamic Management of Water Storage for Flood Control in a Wetland System: A Case Study in Texas. *Water* **2018**, *10*, 325. [CrossRef]
5. Keesstra, S.; Nunes, J.; Novara, A.; Finger, D.; Avelar, D.; Kalantari, Z.; Cerdà, A. The superior effect of nature based solutions in land management for enhancing ecosystem services. *Sci. Total Environ.* **2018**, *610–611*, 997–1009. [CrossRef] [PubMed]
6. Tournebize, J.; Chaumont, C.; Mander, Ü. Implications for constructed wetlands to mitigate nitrate and pesticide pollution in agricultural drained watersheds. *Ecol. Eng.* **2017**, *103*, 415–425. [CrossRef]
7. Barbagallo, S.; Cirelli, G.L.; Marzo, A.; Milani, M.; Toscano, A. Hydraulic behaviour and removal efficiencies of two H-SSF constructed wetlands for wastewater reuse with different operational life. *Water Sci. Technol.* **2011**, *64*, 1032–1039. [CrossRef] [PubMed]

8. Aiello, R.; Cirelli, G.L.; Consoli, S.; Licciardello, F.; Toscano, A. Risk assessment of treated municipal wastewater reuse in Sicily. *Water Sci. Technol.* **2013**, *67*, 89–98. [CrossRef] [PubMed]

9. Barbera, A.C.; Borin, M.; Cirelli, G.L.; Toscano, A.; Maucieri, C. Comparison of carbon balance in Mediterranean pilot constructed wetlands vegetated with different C4 plant species. *Environ. Sci. Pollut. Res. Int.* **2015**, *22*, 2372–2383. [CrossRef] [PubMed]

10. Molari, G.; Milani, M.; Toscano, A.; Borin, M.; Taglioli, G.; Villani, G.; Zema, D.A. Energy characterisation of herbaceous biomasses irrigated with marginal waters. *Biomass Bioenergy* **2014**, *70*, 392–399. [CrossRef]

11. Valenti, F.; Porto, S.M.C.; Chinnici, G.; Cascone, G.; Arcidiacono, C. Assessment of citrus pulp availability for biogas production by using a GIS-based model: The case study of an area in southern Italy. *Chem. Eng. Trans.* **2017**, *58*, 529–534.

12. Kynkaanniemi, P.; Ulen, B.; Torstensson, G.A.; Tonderski, K.S. Phosphorus retention in a newly constructed wetland receiving agricultural tile drainage water. *J. Environ. Qual.* **2013**, *42*, 596–605. [CrossRef] [PubMed]

13. Dipu, S.; Salom Gnana Thanga, V. Heavy metal uptake, its effects on plant biochemistry of wetland (constructed) macrophytes and potential application of the used biomass. *Int. J. Environ. Eng.* **2014**, *6*, 43–54. [CrossRef]

14. Gachango, F.G.; Pedersen, S.M.; Kjaergaard, C. Cost-effectiveness analysis of surface flow constructed wetlands (SFCW) for nutrient reduction in drainage discharge from agricultural fields in Denmark. *Environ. Manag.* **2015**, *56*, 1478–1486. [CrossRef] [PubMed]

15. Reinhardt, M.; Gächter, R.; Wehrli, B.; Müller, B. Phosphorus retention in small constructed wetlands treating agricultural drainage waters. *J. Environ. Qual.* **2005**, *34*, 1251–1259. [CrossRef] [PubMed]

16. Johannesson, K.M.; Tonderski, K.S.; Ehde, P.M.; Weisner, S.E.B. Temporal phosphorus dynamics affecting retention estimates in agricultural constructed wetlands. *Ecol. Eng.* **2017**, *103*, 436–445. [CrossRef]

17. EC (European Commission). *Council Directive Concerning the Protection of Waters against Pollution Caused by Nitrates from Agricultural Sources (91/676/EEC)*; European Commission: Brussels, Belgium, 1991.

18. Tournebize, J.; Chaumont, C.; Fesneau, C.; Guenne, A.; Vincent, B.; Garnier, J.; Mander, Ü. Long-term nitrate removal in a buffering pond-reservoir system receiving water from an agricultural drained catchment. *Ecol. Eng.* **2015**, *80*, 32–45. [CrossRef]

19. Groh, T.A.; Gentry, L.E.; David, M.B. Nitrogen removal and greenhouse gas emissions from constructed wetlands receiving tile drainage water. *J. Environ. Qual.* **2015**, *44*, 1001–1010. [CrossRef] [PubMed]

20. Climatic Tables. Available online: https://www.arpae.it/sim/?osservazioni_e_dati/climatologia (accessed on 23 April 2018). (In Italian)

21. Borin, M.; Milani, M.; Salvato, M.; Toscano, A. Evaluation of *Phragmites australis* (Cav.) Trin. evapotranspiration in Northern and Southern Italy. *Ecol. Eng.* **2011**, *37*, 721–728. [CrossRef]

22. Maddison, M.; Soosaar, K.; Mauring, T.; Mander, Ü. The biomass and nutrient and heavy metal content of cattails and reeds in wastewater treatment wetlands for the production of construction material in Estonia. *Desalination* **2009**, *247*, 121–129. [CrossRef]

23. Zheng, Y.; Wang, X.C.; Ge, Y.; Dzakpasu, M.; Zhao, Y.; Xiong, J. Effects of annual harvesting on plants growth and nutrients removal in surface-flow constructed wetlands in Northwestern China. *Ecol. Eng.* **2015**, *83*, 268–275. [CrossRef]

24. Borin, M.; Tocchetto, D. Five years water and nitrogen balance for a constructed surface flow wetland treating agricultural drainage waters. *Sci. Total Environ.* **2007**, *380*, 38–47. [CrossRef] [PubMed]

25. Ware, G.W.; Whitacre, D.M. *The Pesticide Book*, 6th ed.; Meister Media Worldwide: Willoughby, OH, USA, 2004.

26. Knez, M.; Graham, R.D. The impact of micronutrient deficiencies in agricultural soils and crops on the nutritional health of humans. In *Essentials of Medical Geology, Revised Edition*; Selinus, O., Ed.; Springer: Dordrecht, Germany, 2013; pp. 517–533.

27. Roca, N.; Pazos, M.S.; Bech, J. Background levels of potentially toxic elements in soils: A case study in Catamarca (a semiarid region in Argentina). *Catena* **2012**, *92*, 55–66. [CrossRef]

28. Vymazal, J.; Březinová, T. Accumulation of heavy metals in aboveground biomass of *Phragmites australis* in horizontal flow constructed wetlands for wastewater treatment: A review. *Chem. Eng. J.* **2016**, *290*, 232–242. [CrossRef]

29. Gill, L.W.; Ring, P.; Higgins, N.M.P.; Johnston, P.M. Accumulation of heavy metals in a constructed wetland treating road runoff. *Ecol. Eng.* **2014**, *70*, 133–139. [CrossRef]

30. Angelova, V.R.; Ivanova, R.V.; Todorov, J.M.; Ivanov, K.I. Lead, cadmium, zinc, and copper bioavailability in the soil-plant-animal system in a polluted area. *Sci. World J.* **2010**, *10*, 273–285. [CrossRef] [PubMed]

31. Weiss, J.V.; Emerson, D.; Megonigal, J.P. Geochemical control of microbial Fe (III) reduction potential in wetlands: Comparison of the rhizosphere to non-rhizosphere soil. *FEMS Microbiol. Ecol.* **2004**, *48*, 89–100. [CrossRef] [PubMed]

32. Harguinteguy, C.A.; Fernández Cirelli, A.; Pignata, M.L. Heavy metal accumulation in leaves of aquatic plant *Stuckenia filiformis* and its relationship with sediment and water in the Suquía River (Argentina). *Microchem. J.* **2014**, *114*, 111–118. [CrossRef]

33. D.Lgs. 152 (Decreto legislativo). Norme in materia ambientale. 2006. (In Italian)

34. Carta del Fondo Naturale-Antropico dei Metalli Pesante. Available online: http://ambiente.regione.emilia-romagna.it/geologia/temi/metalli-pesanti/carta-del-fondo-naturale-antropico-della-pianura-emiliano-romagnola-alla-scala1-250-000-2012 (accessed on 23 April 2018). (In Italian)

water

MDPI

Article

Effects of Aeration, Vegetation, and Iron Input on Total P Removal in a Lacustrine Wetland Receiving Agricultural Drainage

Yuanchun Zou [1], Linlin Zhang [1,2,3], Luying Wang [1,3], Sijian Zhang [1,3] and Xiaofei Yu [1,*]

[1] Key Laboratory of Wetland Ecology and Environment & Jilin Provincial Joint Key Laboratory of Changbai Mountain Wetland and Ecology, Northeast Institute of Geography and Agroecology, Chinese Academy of Sciences, Changchun 130102, China; zouyc@iga.ac.cn (Y.Z.); zhanglinlin@iga.ac.cn (L.Z.); wangluying@iga.ac.cn (L.W.); zhangsijian@iga.ac.cn (S.Z.)

[2] The Institute of Geographic Sciences and Natural Resources Research, Chinese Academy of Sciences, Beijing 100101, China

[3] University of Chinese Academy of Sciences, Beijing 100049, China

* Correspondence: yuxf@iga.ac.cn; Tel.: +86-431-8554-2274

Received: 7 December 2017; Accepted: 6 January 2018; Published: 11 January 2018

Abstract: Utilizing natural wetlands to remove phosphorus (P) from agricultural drainage is a feasible approach of protecting receiving waterways from eutrophication. However, few studies have been carried out about how these wetlands, which act as buffer zones of pollutant sinks, can be operated to achieve optimal pollutant removal and cost efficiency. In this study, cores of sediments and water were collected from a lacustrine wetland of Lake Xiaoxingkai region in Northeastern China, to produce a number of lab-scale wetland columns. Ex situ experiments, in a controlled environment, were conducted to study the effects of aeration, vegetation, and iron (Fe) input on the removal of total P (TP) and values of dissolved oxygen (DO) and pH of the water in these columns. The results demonstrated the links between Fe, P and DO levels. The planting of *Glyceria spiculosa* in the wetland columns was found to increase DO and pH values, whereas the Fe:P ratio was found to inversely correlate to the pH values. The TP removal was the highest in aerobic and planted columns. The pattern of temporal variation of TP removals matched first-order exponential growth model, except for under aerobic condition and with Fe:P ratio of 10:1. It was concluded that Fe introduced into a wetland by either surface runoff or agricultural drainage is beneficial for TP removal from the overlying water, especially during the growth season of wetland vegetation.

Keywords: agricultural runoff; phosphorus; iron; pollution control; treatment wetland; Lake Xingkai

1. Introduction

Phosphorus (P), as one of the most important biogenic elements in wetlands, is strongly controlled by the transformation of iron (Fe). The key roles of Fe have been increasingly valued by researchers [1–3]. Under anaerobic conditions in wetlands, microorganisms can obtain energy by oxidizing organic compounds via reducing Fe(III) as an electron acceptor [4,5], and Fe(II) will be re-oxidized by O_2 when the hydrological regime of wetlands shift from wet to unsaturated. During the rotation of this "ferrous–ferric redox wheel", the newly formed Fe(III) (hydr)oxides by the oxidation of O_2 can strongly bind to $PO_4{}^{3-}$ and reduce the total phosphorus (TP) in the water [6–8], a process through which the wetland soils/sediments become P sinks. When the hydrological regime of wetlands shifts from unsaturated to wet, the Fe(III)-P complexes will be re-reduced by anaerobic microorganisms after the depletion of O_2 and $NO_3{}^-$, and the former $PO_4{}^{3-}$ complex will be released again into the soil/sediment pore water and overlying water, a process through which sediments/soils become P sources [9].

Although ferrous or ferric solutions have been used in wastewater treatment to remove P for many years, there are not many studies on P demobilization in natural wetlands [10]. Some case studies in reservoirs and lakes confirmed the effectiveness of ferric solutions in removing P from overlying water [11–15]. A Fe:P ratio has also been used to predict the P adsorption capacity of soils/sediments due to the significant negative correlation between dissolved P and Fe:P [16,17]. Considering the risk of dissolved Fe-P complexes under anaerobic conditions over the long term, artificial aeration has been implemented to avoid this risk and increase the P removal efficiency [18]. In addition to artificial aeration, aeration through wetland vascular plants is also an important natural aeration mechanism [19–21].

Therefore, there is a coupled relationship between Fe, O, and P in wetlands, which affects the level of P in a wetland and further affects the eutrophication of water bodies; however, it is unclear how the exogenous Fe inputs from natural runoff and/or agricultural drainage affect the transport and transformation of P in wetlands when redox conditions change with artificial or natural aeration.

In this study, we compared the effects of aeration and rooting plants on dissolved O_2, pH, and TP removal under a lower Fe:P ratio of 5 and a higher ratio of 10 in the overlying water. The objectives of this study were (1) to test the temporal responses of O_2, pH, and TP in the overlying water to different redox conditions; and (2) to assess TP removal from the overlying water and test the effectiveness of Fe inputs on TP removal in the wetlands receiving agricultural drainage.

2. Materials and Methods

2.1. Sampling Site and Soil Core Sampling

Lake Xingkai is a transboundary lake shared by China and Russia. A narrow sandy ridge on the northern shore is separated from Lake Xingkai and forms a smaller lake named Xiaoxingkai that belongs to China. At Lake Xingkai, the annual mean precipitation is 561 mm, and the mean temperature is 3.5 °C [22]. The Fe concentration of the lake water can be as high as 0.62 mg·L^{-1} and is the highest concentration found among the waters in the Heilongjiang River system [23]. This high concentration is caused by both the high background levels of natural runoff and the agricultural drainage from upstream [24]. The sediment cores were collected from a lacustrine wetland covered by *Glyceria spiculosa* communities in Lake Xiaoxingkai (45°13'47" N, 132°46'26" E). *G. spiculosa* usually develops aerenchyma in its stems, leaves, rhizomes, and roots, which make this species adaptive to submerged conditions, and it has become one of the dominant species in lacustrine wetlands around Lake Xingkai. The sampled lacustrine wetland is mainly supplied by lake water and precipitation and has periodically received agricultural drainage.

Twenty-four intact wetland soil cores were collected randomly using polyvinyl chloride (PVC) tubes (30 cm length × 6.8 cm internal diameter). Each core included 15 cm of submerged soil and 6 cm of water and was sealed with plastic bungs, stored in a portable cold closet and then transported to the laboratory within 48 h. In addition, representative and homogeneous seedlings of *G. spiculosa* were collected and transferred to the laboratory as well.

2.2. Experimental Design and Chemical Analyses

The incubation was designed with three treatment factors and two levels (aeration × plant × Fe:P × replicates), and each treatment had three replicates. To not alter the integrity of the soil column, the overlying water of the 12 cores randomly selected from the 24 cores was siphoned out, and 48 seedlings (four seedlings per core) of *G. spiculosa* with the same height (approximately 7 cm) were selected for careful implantation in these cores using long forces. After three days of successful introductions and when all the plants survived, the overlying water siphoned out was siphoned back into the soil column. Six randomly selected cores with plants and six without plants were dosed and adjusted so that the overlying water contained 5 mg·L^{-1} Fe and 1 mg·L^{-1} P (Fe:P = 5) with FeCl$_2$·4H$_2$O and 1 mg·L^{-1} NaH$_2$PO$_4$·2H$_2$O solutions, respectively. The rest of the cores with or without plants were dosed and adjusted to contain 10 mg·L^{-1} Fe and 1 mg·L^{-1} P (Fe:P = 10). The aeration treatments were adjusted to establish high (>6 mg·L^{-1}, aerobic) and low (<2 mg·L^{-1}, anaerobic) DO concentrations by continuous gentle bubbling with O$_2$ or N$_2$, respectively.

Before incubation, the overlying water depth of each core was adjusted to 6 cm. The indoor incubation was completed at an ambient temperature of 23–25 °C for 25 days, with the necessary light compensation provided by a plant growth lamp. The overlying water was monitored and measured seven times on 0, 3, 5, 8, 12, 17, and 24 d. The pH and DO were determined in situ using the EXO2 Multiparameter Sonde (YSI Incorp., Yellow Springs, OH, USA). The TP was determined by extracting 15 mL of water with a syringe, and then this water was replaced with 15 mL of deionized water. The TP in the extracted water sample was measured using the ammonium molybdate spectrophotometric method (UV 2550, Shimadzu, Japan) after digestion with HClO$_4$-H$_2$SO$_4$ for 0.5 h [25].

2.3. Data and Statistical Analyses

TP removal percentages (R_{TP}) were calculated according to the formula

$$R_{TP} = \frac{C_0 - C_i}{C_0} \times 100\% \tag{1}$$

where C_0 and C_i are the TP (mg·L^{-1}) at the initial and the *i*th sampling, respectively, and *i* is the sampling frequency.

Two-way repeated measures analysis of variance (ANOVA) were performed to test the main and interaction effects of aeration and plant treatments using SPSS Statistics 21.0 (SPSS Inc., Chicago, IL, USA). All the means and standard errors were calculated using Origin Pro 8.0 (OriginLab Corp., Northampton, UK), and all the graphics were drawn using Origin Pro 8.0.

3. Results

The temporal changes of DO, pH, and TP removal varied with the different treatments. The two-way repeated measures ANOVA showed that the main and interaction effects of the aeration and plant treatments on the DO were all significant under the lower Fe:P ratio (Fe:P = 5), while only the main effects of the aeration and plant treatments on the pH were significant. For TP removal, the main effect of aeration was extremely significant and even caused the interaction effect to be significant as well (Table 1).

Table 1. Results of two-way repeated measures analysis of variance when Fe:P = 5. DO, dissolved oxygen; TP, total phosphorus.

Variable	Time		Time × Plant		Time × Aeration		Time × Plant × Aeration	
	F	*P*	*F*	*P*	*F*	*P*	*F*	*P*
DO	32.728	<0.0001	2.488	0.043	32.652	<0.0001	2.538	0.038
pH	21.401	<0.0001	2.658	0.026	7.001	<0.0001	1.920	0.097
TP removal	283.922	<0.0001	1.437	0.220	247.775	<0.0001	3.444	0.007

When more Fe was introduced (Fe:P = 10), the main effects of the aeration and plant treatments on the DO were both significant, while the interaction effect was non-significant. For pH, only the main effect of aeration was significant. For TP removal, neither the main effects nor the interaction effect were significant (Table 2).

Table 2. Results of two-way repeated measures analysis of variance when Fe:P = 10.

Variable	Time		Time × Plant		Time × Aeration		Time × Plant × Aeration	
	F	*P*	*F*	*P*	*F*	*P*	*F*	*P*
DO	48.639	<0.0001	4.412	0.001	34.669	<0.0001	1.377	0.243
pH	23.570	<0.0001	2.057	0.076	14.717	<0.0001	1.940	0.093
TP removal	6.309	<0.0001	2.154	0.064	1.751	0.130	2.191	0.060

3.1. DO Variation

The DO of the overlying water in the aerobic cores increased with the incubation time from approximately 6.0 to 9.0 mg·L^{-1}, while the DO in the anaerobic cores kept stable at approximately 1.0 mg·L^{-1}. The former was much higher than the latter, and the differences between these two values increased with the incubation time (Figure 1). In the first five days after beginning the incubation, the differences between the plant and non-plant cores were non-significant; however, significantly higher DO levels were observed in the plant cores than the non-plant cores except under anaerobic conditions with a higher Fe:P ratio after five days (Figure 1b).

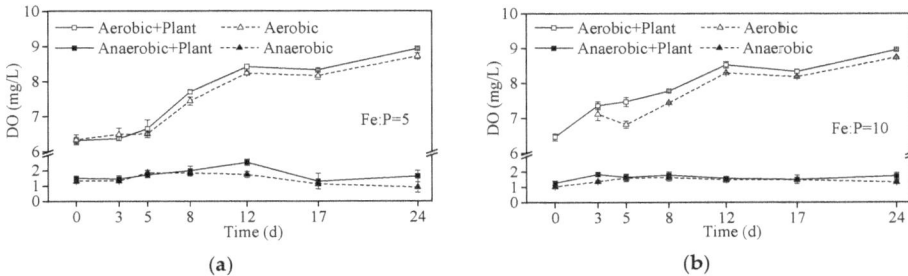

Figure 1. Variation of dissolved oxygen (DO) in the overlying water. (**a**) Iron: phosphorus ratio (Fe:P) = 5; (**b**) Fe:P = 10. Error bars represent the standard errors.

3.2. pH Variation

All the pH values of the overlying water in the different treatments fluctuated during the incubation. Compared with the other three treatments, the pH values of the aerobic and plant cores were the highest, for both Fe:P ratios. In the anaerobic cores, the pH values decreased compared with those in the beginning of the incubation. Compared with the pH values under two Fe:P ratios, the higher pH could be observed in the cores with the lower Fe:P ratio (Figure 2). The mean pH decreased from 6.93 to 6.92 for aerobic and plant cores, from 6.58 to 6.14 for aerobic cores, from 6.56 to 5.78 for anaerobic and plant cores, and from 6.43 to 5.95 for anaerobic cores. The mean pH values of the plant cores were greater than those of the non-plant cores, although this difference at higher Fe:P ratios was non-significant ($P = 0.76$).

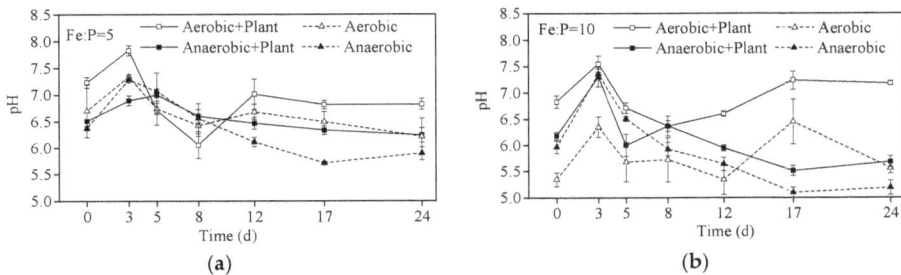

Figure 2. Variation of pH in the overlying water. (**a**) Fe:P = 5; (**b**) Fe:P = 10.

3.3. TP Removal Variation

Under the lower Fe:P ratio, two types of TP removal curves could be observed. For the aerobic cores, TP removal fluctuated at approximately the 95% level. For the anaerobic cores, TP removal increased exponentially. After three days of rapid growth, TP removal in the plant cores and non-plant cores rapidly increased by 2.5 times and 3.3 times more than the initial levels, respectively (Figure 3a). Under the higher Fe:P ratio, all TP removals fluctuated with the extension of incubation time but increased in comparison with the levels in the beginning (Figure 3b). Among the various treatments, TP removals were the highest in the aerobic and plant cores than in the other treatments. Except for the aerobic and plant cores under the higher Fe:P ratio, the TP removal curves could be fitted by first-order exponential growth equations (Tables 3 and 4).

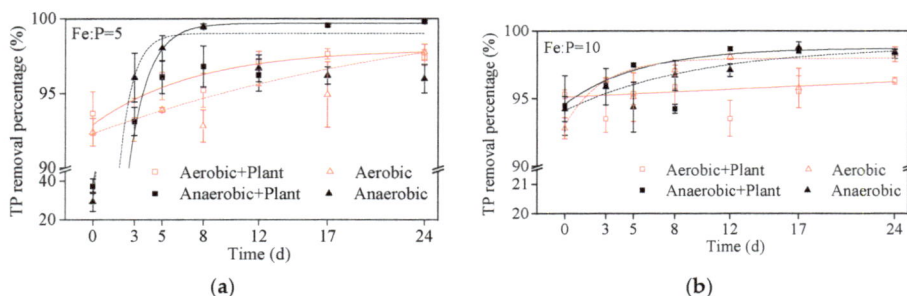

Figure 3. Removal percentage of total phosphorus (TP) in the overlying water. (**a**) Fe:P = 5; (**b**) Fe:P = 10.

Table 3. Fitting equations of TP removal when Fe:P = 5.

Treatment	Fitting Equation	R^2	P
Aerobic + Plant	$y = -5.01 \times \exp(-x/7.37) + 97.90$	0.75	<0.001
Aerobic	$y = -9.16 \times \exp(-x/26.94) + 101.44$	0.73	<0.001
Anaerobic + Plant	$y = -61.76 \times \exp(-x/1.40) + 99.68$	0.90	<0.001
Anaerobic	$y = -69.73 \times \exp(-x/0.94) + 99.00$	0.71	<0.001

Table 4. Fitting equations of TP removal when Fe:P = 10.

Treatment	Fitting Equation	R^2	P
Aerobic + Plant	$y = 0.046x + 95.08$ [1]	0.46	0.056
Aerobic	$y = -4.91 \times \exp(-x/2.72) + 97.92$	0.85	<0.001
Anaerobic + Plant	$y = -4.13 \times \exp(-x/5.50) + 98.73$	0.14	<0.001
Anaerobic	$y = -4.76 \times \exp(-x/9.77) + 98.88$	0.81	<0.001

Note: [1] First-order exponential equation was not applied for this treatment for the parameters were not reasonable and $R^2 < 0.5$.

4. Discussion

4.1. Interactions among Aeration, Plants, and Fe:P

Fe, P, O, and H are transformed and coupled to each other in wetland environments [5], and their interactions occur not only at an abiotic level but also a biotic level [4–7]. Regardless of bubbling with either O_2 or N_2, the DO in the plant cores was higher than that in the non-plant cores, indicating that the existence of wetland plants plays an important role in increasing DO concentrations in overlying water (Figure 1, Tables 1 and 2), and this result confirmed the process of internal aeration in wetland plants through aerenchyma [19,20]. Considering the *G. spiculosa* seedlings used in this study were newly planted, their roots were not well-developed, and the internal transport capacity of O_2 was restricted. However, when roots grow well and densely in wetland habitats in the field, natural aeration would be effective and create oxic conditions even though the soils are submerged during the growing season. When plant vitality is not strong and agricultural drainage flows into wetlands, artificial aeration is recommended. The effectiveness of aeration on P demobilization in wetland soils, however, depends on the Fe content [16,17] and the type of Fe-P complexes [18].

It is interesting that the pH values of the plant cores were greater than those of the non-plant cores, regardless of the O_2 levels, indicating that the presence of plants could increase the pH of water bodies (Figure 2). More O_2 released into the water by radial loss from roots would oxidize ferrous ions to ferric ions, and OH^- would be produced during this process. The newly formed ferric ions would be subject to hydrolysis. The higher Fe:P ratio could also decrease the pH (Figure 2), which could be attributed to the increased hydrolysis of Fe ions and the extra release of H^+, resulting in acidification [10,26]. Theoretically, the actual O_2, pH, and Fe:P in wetlands would be the balance among these coupled processes. Therefore, the Fe introduced by agricultural drainage and its transformation with P in soils should be further studied in the field to test the effectiveness of TP removal under different aeration conditions.

4.2. TP Removals under Different Fe:P Conditions and Management Recommendations

P is usually limited in most natural wetlands and other aquatic ecosystems; however, excessive P input has become one of the key control factors causing water eutrophication [1,3]. Among the exogenous P inputs, agricultural drainage is one of the most important sources in Lake Xiaoxingkai. Although there is no survey of how many fertilizers and pesticides have been used in the farms around the lake, the total amount of fertilizers and pesticides used in Heilongjiang Province where the lake is located, as well as the amount used per hectare, showed logistic growth from 1996 to 2015, and a considerable portion of the unused fertilizers and pesticides eventually enter and sink into the wetlands downstream [27]. For Lake Xiaoxingkai, over 0.5 billion m^3 of lake water were pumped for irrigation of the paddy fields in the state farms adjacent to Lake Xingkai, and large amounts of agricultural drainage return to the remaining natural lacustrine wetlands during the drainage return flow period. Consequently, P, Fe, and sediments have accumulated in these wetlands [24]. According to a recent survey, Lake Xiaoxingkai was experiencing moderate eutrophication, and P was identified as the restriction element [28]; therefore, the agricultural drainage that flows and accumulates in these wetlands with Fe:P should be considered as the key factor to controlling eutrophication internally.

TP removal under different Fe:P ratios gradually increased with incubation time (Tables 3 and 4), indicating that the input of Fe reduced TP to a certain extent. According to the fitting equations (Tables 3 and 4), although all the final TP removals would be more than 97.9% when the incubation time was long enough, the initial TP removals were 92.89, 92.28, 37.92, and 29.27% for the different treatments under the lower Fe:P ratio, and 95.08, 93.01, 94.6, and 94.12% for the higher Fe:P ratio. These results suggested that the higher Fe:P ratio could increase initial TP removal as in the field observations by Kleeberg et al. [15], especially under anaerobic conditions (increased TP removal by 56.68% for the anaerobic and plant cores and by 64.85% for the anaerobic cores).

In addition, when the Fe:P ratio is as high as 10 in the wetland, our results (Figure 3b) showed that the rapid growth period of TP removal was significantly shortened compared with the lower Fe:P ratio during the first three days (Figure 3a). This result indicated that more Fe could compensate for the disadvantage of low TP removal when the treatment time is too short, and there are a lack of aeration measures. Therefore, more Fe introduced by surface runoff, groundwater, or agricultural drainage could be beneficial to increasing TP removal from the overlying water, especially where wetland plants are growing well and are dense. For the practical use of Fe introduction method in natural water bodies' restoration, the Fe species and dosage should be further specified according to the local P and DO concentrations and an Fe:P ratio of 10 is recommended.

5. Conclusions

During indoor wetland incubation, the DO in the overlying water was affected both by the aeration and plant treatment and plant increased DO. The pH was affected by the aeration and plant treatment as well. TP removal was only affected by aeration under the lower Fe:P ratio. Introducing more Fe could increase initial TP removal and shorten the initial rapid growth period, especially under anaerobic conditions. We concluded that greater Fe introduction by agricultural drainage could be beneficial to increasing TP removal from the overlying water. This study contributes to highlighting the importance of further understanding the elemental coupled cycling involved in the water purification function of in wetland ecosystem and providing a scientific basis for the protection and management of Lake Xiaoxingkai. Considering the complicated interactions, more studies focusing on Fe and P fluxes introduced by agricultural drainage and their coupled transformation should be carried out in the field under different hydrological regimes and redox conditions, through which more scientifically reasonable and cost-effective management strategies and techniques could be developed.

Acknowledgments: The authors express gratitude to the reviewers and editors for their critical comments on an earlier version of the manuscript. This work was supported by the National Key Research & Development Program of China (2016YFC0500408), the National Natural Science Foundation of China (41271107, 41471079, 41771120), and the Northeast Institute of Geography and Agroecology, CAS (IGA-135-05).

Author Contributions: Y.Z. and X.Y. conceived and designed the experiments; L.Z. performed the experiments; L.W. and analyzed the data; S.Z. contributed reagents and materials; G.S. edited the manuscript prior to submission; Y.Z. wrote the paper.

Conflicts of Interest: The authors declare no conflict of interest.

References

1. Edgar, F.L.; Keenan, L.W. Managing phosphorus-based, cultural eutrophication in wetlands: A conceptual approach. *Ecol. Eng.* **1997**, *9*, 109–118.
2. Dunne, E.J.; Reddy, R.; Clark, M.W. Biogeochemical indices of phosphorus retention and release by wetland soils and adjacent stream sediments. *Wetlands* **2006**, *26*, 1026–1041. [CrossRef]
3. Zamparas, M.; Zacharias, I. Restoration of eutrophic freshwater by managing internal nutrient loads: A review. *Sci. Total Environ.* **2014**, *496*, 551–562. [CrossRef] [PubMed]
4. David, E.; Eric, R.; Twining, B.S. The microbial ferrous wheel: Iron cycling in terrestrial, freshwater, and marine environments. *Front. Microbiol.* **2012**, *3*, 383.
5. Li, Y.C.; Yu, S.; Strong, J.; Wang, H.L. Are the biogeochemical cycles of carbon, nitrogen, sulfur, and phosphorus driven by the "FeIII–FeII redox wheel" in dynamic redox environments? *J. Soils Sediments* **2012**, *12*, 683–693. [CrossRef]
6. Roden, E.E.; Wetzel, R.G. Kinetics of microbial Fe (III) oxide reduction in freshwater wetland sediments. *Limnol. Oceanogr.* **2002**, *47*, 198–211. [CrossRef]
7. Weber, K.A.; Achenbach, L.A.; Coates, J.D. Microorganisms pumping iron: Anaerobic microbial iron oxidation and reduction. *Nat. Rev. Microbiol.* **2001**, *4*, 752–764. [CrossRef] [PubMed]
8. Gunnars, A.; Blomqvist, S.; Johansson, P.; Andersson, C. Formation of Fe (III) oxyhydroxide colloids in freshwater and brackish seawater, with incorporation of phosphate and calcium. *Geochim. Cosmochim. Acta* **2002**, *66*, 745–758. [CrossRef]
9. Qualls, R.G.; Sherwood, L.J.; Richardson, C.J. Effect of natural dissolved organic carbon on phosphate removal by ferric chloride and aluminum sulfate treatment of wetland waters. *Water Resour. Res.* **2009**, *45*, W09414. [CrossRef]
10. Zou, Y.C.; Grace, M.R.; Roberts, K.L.; Yu, X.F. Thin ferrihydrite sediment capping sequestrates phosphorus experiencing redox conditions in a shallow temperate lacustrine wetland. *Chemosphere* **2017**, *185*, 673–680. [CrossRef] [PubMed]
11. Cooke, G.D.; Welch, E.B.; Martin, A.B.; Fulmer, D.G.; Hyde, J.B.; Schrieve, G.D. Effectiveness of Al, Ca, and Fe salts for control of internal phosphorus loading in shallow deep lakes. *Hydrobiologia* **1993**, *253*, 323–335. [CrossRef]
12. Smolders, A.J.P.; Lamers, L.P.M.; Moonen, M.; Zwaga, K.; Roelofs, J.G.M. Controlling phosphate release from phosphate-enriched sediments by adding various iron components. *Biogeochemistry* **2001**, *54*, 219–228. [CrossRef]
13. Sherwood, L.J.; Qualls, R. Stability of phosphorus within a wetland soil following ferric chloride treatment to control eutrophication. *Environ. Sci. Technol.* **2001**, *35*, 4126–4131. [CrossRef] [PubMed]

14. Deppe, T.; Benndorf, J. Phosphorus reduction in a shallow hypereutrophic reservoir by in-lake dosage of ferrous iron. *Water Res.* **2002**, *36*, 4525–4534. [CrossRef]

15. Kleeberg, A.; Herzog, C.; Hupfer, M. Redox sensitivity of iron in phosphorus binding does not impede lake restoration. *Water Res.* **2013**, *47*, 1491–1502. [CrossRef] [PubMed]

16. Jensen, H.S.; Kristensen, P.; Jeppesen, E.; Skytthe, A. Iron-phosphorus ratio in surface sediment as an indicator of phosphate release from aerobic sediments in shallow lakes. *Hydrobiologia* **1992**, *235*, 731–743. [CrossRef]

17. Grace, M.R.; Scicluna, T.R.; Vithana, C.L.; Symes, P.; Lansdown, K.P. Biogeochemistry and cyanobacterial blooms: Investigating the relationship in a shallow, polymictic, temperate lake. *Environ. Chem.* **2010**, *7*, 443–456. [CrossRef]

18. Gächter, R.; Müller, B. Why the phosphorus retention of lakes does not necessarily depend on the oxygen supply to their sediment surface. *Limnol. Oceanogr.* **2003**, *48*, 929–933. [CrossRef]

19. Colmer, T.D. Long-distance transport of gases in plants: A perspective on internal aeration and radial oxygen loss from roots. *Plant Cell Environ.* **2003**, *26*, 17–36. [CrossRef]

20. Takahashi, H.; Yamauchi, T.; Colmer, T.D.; Nakazono, M. Aerenchyma formation in plants. *Plant Cell Monogr.* **2014**, *21*, 247–265.

21. Zhang, Q.; Huber, H.; Beljaars, S.J.M.; Birnbaum, D.; De Best, S.; De Kroon, H.; Visser, E.J.W. Benefits of flooding-induced aquatic adventitious roots depend on the duration of submergence: Linking plant performance to root functioning. *Ann. Bot.* **2017**, *120*, 171–180. [CrossRef] [PubMed]

22. Zou, Y.C.; Wang, G.P.; Grace, M.R.; Lou, X.N.; Yu, X.F.; Lu, X.G. Response of two dominant boreal freshwater wetland plants to manipulated warming and altered precipitation. *PLoS ONE* **2014**, *9*, e104454. [CrossRef] [PubMed]

23. Lu, L.; Dong, C.Z.; Zhao, C.X.; Liu, Y.; Zhan, P.R. Physio-chemical characteristics of different waters in Heilongjing System. *J. Fish. China* **2003**, *27*, 364–370.

24. Wang, G.D.; Wang, M.; Yuan, Y.X.; Lu, X.G.; Jiang, M. Effects of sediment load on the seed bank and vegetation of calamagrostis angustifolia, wetland community in the national natural wetland reserve of lake xingkai, China. *Ecol. Eng.* **2014**, *63*, 27–33. [CrossRef]

25. Xie, X.Q.; Wang, L.J. *Observation and Analysis of Water Environment Factors (Standard Methods for Observation and Analysis in Chinese Ecosystem Research Networks)*; China Standard Press: Beijing, China, 1998; pp. 56–89.

26. Flynn, C.M., Jr. Hydrolysis of inorganic iron (III) salts. *Chem. Rev.* **1984**, *84*, 31–41. [CrossRef]

27. Zou, Y.; Wang, L.; Xue, Z.; Mingju, E.; Jiang, M.; Lu, X.; Yang, S.; Shen, X.; Liu, Z.; Sun, G.; et al. Impacts of agricultural and reclamation practices on wetlands in the Amur River Basin. *Wetlands* **2018**, in press. [CrossRef]

28. Yu, S.L.; Li, X.J.; Chen, G.S.; Zhang, J.T.; Yang, Y.L.; Yan, Y.; Lu, X.R.; Zhang, C.Y. Analysis of eutrophication and terrestrialization of Xiaoxingkai Lake. *Wetl. Sci.* **2016**, *14*, 271–275.

Article

Performance of Iron Plaque of Wetland Plants for Regulating Iron, Manganese, and Phosphorus from Agricultural Drainage Water

Xueying Jia [1,2], Marinus L. Otte [3], Ying Liu [1,2], Lei Qin [1,2], Xue Tian [1], Xianguo Lu [1,4], Ming Jiang [1,4,*] and Yuanchun Zou [1,4,*]

[1]   Key Laboratory of Wetland Ecology and Environment, Northeast Institute of Geography and Agroecology, Chinese Academy of Sciences, Changchun 130102, China; jiaxueying13@mails.ucas.ac.cn (X.J.); liuying2011@163.com (Y.L.); qinlei@iga.ac.cn (L.Q.); tianxue2324@163.com (X.T.); luxg@neigae.ac.cn (X.L.)
[2]   University of Chinese Academy of Sciences, Beijing 100049, China
[3]   Wet Ecosystem Research Group, Department of Biological Sciences, North Dakota State University, Fargo, ND 58105-6050, USA; marinus.otte@ndsu.edu
[4]   Jilin Provincial Joint Key Laboratory of Changbai Mountain Wetland and Ecology, Changchun 130102, China
*   Correspondence: jiangm@iga.ac.cn (M.J.); zouyc@iga.ac.cn (Y.Z.); Tel.: +86-8554-2207 (M.J.)

Received: 30 November 2017; Accepted: 5 January 2018; Published: 8 January 2018

Abstract: Agricultural drainage water continues to impact watersheds and their receiving water bodies. One approach to mitigate this problem is to use surrounding natural wetlands. Our objectives were to determine the effect of iron (Fe)-rich groundwater on phosphorus (P) removal and nutrient absorption by the utilization of the iron plaque on the root surface of *Glyceria spiculosa* (Fr. Schmidt.) Rosh. The experiment was comprised of two main factors with three regimes: Fe^{2+} (0, 1, 20, 100, 500 mg·L^{-1}) and P (0.01, 0.1, 0.5 mg·L^{-1}). The deposition and structure of iron plaque was examined through a scanning electron microscope and energy-dispersive X-ray analyzer. Iron could, however, also impose toxic effects on the biota. We therefore provide the scanning electron microscopy (SEM) on iron plaques, showing the essential elements were iron (Fe), oxygen (O), aluminum (Al), manganese (Mn), P, and sulphur (S). Results showed that (1) Iron plaque increased with increasing Fe^{2+} supply, and P-deficiency promoted its formation; (2) Depending on the amount of iron plaque on roots, nutrient uptake was enhanced at low levels, but at higher levels, it inhibited element accumulation and translocation; (3) The absorption of manganese was particularly affected by iron plague, which also enhanced phosphorus uptake until the external iron concentration exceeded 100 mg·L^{-1}. Therefore, the presence of iron plaque on the root surface would increase the uptake of P, which depends on the concentration of iron-rich groundwater.

Keywords: wetlands; agricultural drainage water; iron plague; phosphorus; manganese

1. Introduction

Agricultural activities are vital to the crop production and economy of terrestrial ecosystems; meanwhile, its practice continues to place environmental pressures on nearby lake and wetland ecosystems [1]. Agricultural drainage water originating from groundwater contains large amount of farm nutrients and high concentration of iron (Fe), which posing an environmental hazard to the surrounding ecosystems. In recent years, the area of rice cultivation has been increased to 4,000,000 ha of land in Heilongjiang Province, China. Such a large-scale agricultural production would inevitably bring great impacts on the surrounding ecosystems, especially for the sustainability and water security of the lake and wetlands.

Wetlands play important roles in nutrient recycling, increasing dissolved O_2 in the rhizosphere, recharging of aquifers, stabilization of water currents, improvement of water quality, and remediation of wastewater (including removal of excess iron, manganese, and other metals) [2–5]. Wetland plants have the capacity to improve water quality through mobilization and absorption of nutrients and contaminants [6]. O_2 diffuses from the roots to their surroundings and oxidizes Fe^{2+} in the rhizosphere into oxyhydroxide iron (Fe plaque) on the root surface of plants [7]. Iron plaque on roots absorbs metals onto its large surface area and co-precipitates with nutrients, such as P. Root activities influence plaque formation by releasing O_2 and exudates, and influencing the rates of enzyme and radial oxygen loss (ROL). Iron plaque formation occurs commonly on the surface of roots of wetland species, including *Typha latifolia* L., *Oryza sativa* L., and *Phragmites communis* Trin. [8].

Metal and nutrient enrichment and deposition in iron plaque are known to occur, but the effects of iron plaque on uptake and translocation of elements are still unclear [9–11]. Some studies reported that the formation of Fe plaque prevents the excessive uptake of Fe, Mn, Zn [12,13], and phosphorus (P) [14]. On the other hand, iron plaque acts as a buffer for P, Zn, Cu, Se, and As. While the plants may be lack nutrients, they are also at risk of metal toxicity [15,16]. The effects of Fe plaque on mineral nutrients and other elements may be related to the amount of deposition on the root. It has been reported that Fe plaque enhanced Zn and P uptake, and could be considered as a nutrient reservoir, but may also act as a barrier when excess Fe plaque forms on the root surface [7,17]. The presence of P is important in energy metabolism, the biosynthesis of nucleic acids, photosynthesis, and enzyme regulation [18], and also plays an important role in biogeochemical cycling with Fe and Mn [19,20]. The studies have demonstrated that P-deficiency may enhance the formation of iron plaque on the roots of plants [21,22]. At the same time, iron plaque on the root surface may affect the uptake and translocation of P, Mn, Zn, Cu, As, and Cd. Using reactive materials to immobilize P in soils or manure is considered a new way of managing P eutrophication [23].

In order to adapt to long-term iron-rich environmental changes, wetland plants have different growth strategies in different environments. Higher plants have developed two divided strategies to acquire slightly soluble iron from the rhizosphere: the chelation strategy of graminaceous plants and the reduction strategy of nongraminaceous plants [9]. The "exclusion" strategy is supported by studies, which showed that some species exclude phytotoxic metals rather than absorbing them. The presence of iron plaque on the root surface is another strategy, which is an adaptation to flooding. With the iron plaque strategy, the absorption and translocation of a large number of elements will be affected.

The aim of this study was to determine the effects of iron toxicity on plants when they are grown in the iron-rich groundwater and agricultural drainage water recharge marsh, such as the Xiaoxingkai Lake wetland in the south of the Sanjiang Plain, China. Xiaoxingkai Lake is a Chinese National Nature Reserve and a boundary lake between China and Russia. The experimental wetland plant was *Glyceria spiculosa* (*G. spiculosa*), a typical dominant but sensitive species with developed aerenchyma in the roots. Therefore, the variation of *G. spiculosa* has a certain indicative function on the change of the surrounding habitat. In the present investigation, our main objectives were to study the effects of: (a) P and Fe concentrations on the formation of iron plaque on root surfaces; (b) iron plaque on the uptake and translocation of Fe, Mn, and P; and (c) iron-rich groundwater as a reactive resource to manage P eutrophication water caused by the agricultural drainage water.

2. Materials and Methods

2.1. Greenhouse Experiment

To investigate the influence of various Fe^{2+} and P concentrations levels on the formation of iron plaque and its function on elements uptake and translocation, an experiment was performed with *G. spiculosa*. The experiment was carried out in a greenhouse at Northeast Institute of Geography and Agroecology, Chinese Academy of Sciences (43°59′53″ N and 125°23′48″ E) during July to

September, 2016. Average air temperatures during the day (7:00–19:00) were 25–28 °C, and during the night they were (19:00–7:00) 15–20 °C.

Before beginning the experiment, tillers training were pre-cultured for two weeks in the greenhouse. The plants were tiled on the plastic vessels (100 × 60 × 5 cm) containing 10 L half-strength Hoagland nutrient solution (Hoagland and Arnon 1950), and the solution was renewed every seven days. Tillers of uniform size (15–18 cm) were selected on the rhizomes of the *G. spiculosa*, cleaned with deionized water, and transplanted to plastic pots (12 cm diameter, 15 cm height) into quartz sand and half-strength Hoagland nutrient solution for one week. The Hoagland solution was composed of 0.5 mM $NH_4H_2PO_4$, 2 mM $Ca(NO_3)_2 \cdot 4H_2O$, 3 mM KNO_3, 1 mM $MgSO_4 \cdot 7 H_2O$, 4.57 μM $MnCl_2 \cdot 4H_2O$, 23.13 μM H_3BO_3, 0.382 μM $ZnSO_4 \cdot 7H_2O$, 0.16 μM $CuSO_4 \cdot 5H_2O$, and 0.0695 μMMoO_3.

2.2. Experiment Design

The experiment was comprised of two main factors with three regimes: Fe^{2+} (0, 1, 20, 100, 500 mg·L^{-1}) and P (0.01, 0.1, 0.5 mg·L^{-1}) (Table 1). The first regime lasted three weeks, with ferrous sulfate in the half-strength Hoagland nutrient solution, which was designed to examine the influence of different Fe^{2+} concentrations and the resistance threshold of wetland plants to excessive Fe^{2+} treatment. The second regime lasted one week, with ferrous sulfate addition, but did not contain phosphorus in the solution, as it intended to flush out residual PO_4-P and prompt the iron plaque formation on the roots. The third regime lasted two weeks with ammonium dihydrogen phosphate addition, to determine the influence of iron plague in the asorption and translocation of Fe, Mn, and P. Each treatment consisted of five replications in a large rectangular trough (67 × 26 × 22 cm), with a total of 75 pots. The pots were waterlogged (2–3 cm) with the testing solution, and the bottom of each pot had five small holes. Each of the pots were wrapped with a black plastic bag. The nutrient solution was renewed every seven days, and the pH was adjusted to 5.5 using 0.2 mol·L^{-1} NaOH or HCl every two days.

2.3. Scanning Electron Microscopy (SEM) and Energy Dispersive X-Ray Spectrometry (EDS)

Root material was washed with tap water, then rinsed three times in deionized water. For the scanning electron microscopy analysis, 1 cm long lateral fresh root samples of *G. spiculosa* were prepared and fixed in formyl acetic alcohol (FAA). Samples were dehydrated in ethyl series and dried using liquid CO_2 (CPD 030 model, Bal-Tec Co., Balzers, Liechtenstein). The morphology of roots was observed using a scanning electron microscopy (SS-550, Shimadzu, Tokyo, Japan).

2.4. Chemical Analysis

At the end of the experiment, the plants were harvested and separated into roots, stems, and leaves, and washed with deionised water three times. Fresh root samples were weighted 1 g and then soaked by dithionite-citrate-bicarbonate (DCB) [24]. The DCB solution consisted of 40 mL of 0.3 M sodium dithionite and 5 mL of 1 M sodium citrate with 3.0 g sodium dithionite. The roots in the solution were shaken for 3 h for a complete extraction. The extrated roots were dried at 60 °C to constant weight. Fe concentrations in DCB were determined by atomic absorption (Optima 2000DV, Perkin Elmer, Waltham, MA, USA).

The other plant biomass comparments were dried at 55 °C for 72 h, used to measure iron, manganese, and phosphrous with perchloric ($HClO_4$) acids and nitric (HNO_3) (1:5, *v/v*). The concentrations of iron and manganese in plant tissues were assayed by atomic absorption spectrometry (Optima 2000DV, Perkin Elmer, Waltham, MA, USA), and phophrous was determined with Molybdenum antimony colorimetric method by Automatic chemistry analyzer (Smartchem 300, Advanced Monolithic Systems, Graz, Italy).

Table 1. Experimental treatment for a fully factorial design with three regimes and two main factors: Fe^{2+} (0, 1, 20,100, 500 mg·L^{-1} Fe) and P (0.01, 0.1, 0.5 mg·L^{-1} P).

Time (Weeks)	Treatments				
	Fe 0	**Fe 1**	**Fe 20**	**Fe 100**	**Fe 500**
1	Fe 0 + H	Fe 1 + H	Fe 20 + H	Fe 100 + H	Fe 500 + H
2	Fe 0 + H	Fe 1 + H	Fe 20 + H	Fe 100 + H	Fe 500 + H
3	Fe 0 + H	Fe 1 + H	Fe 20 + H	Fe 100 + H	Fe 500 + H
4	Fe 0+ H(−P)	Fe 1 + H(−P)	Fe 20+ H(−P)	Fe 100+ H(−P)	Fe 500+ H(−P)
5	+0.01 PH +0.1 PH +0.5 PH	+0.01 PH +0.1 PH +0.5 PH	+0.01 PH +0.1 PH +0.5 PH	+0.01 PH +0.1 PH +0.5 PH	+0.01 PH +0.1 PH +0.5 PH
6	+0.01 PH +0.1 PH +0.5 PH	+0.01 PH +0.1 PH +0.5 PH	+0.01 PH +0.1 PH +0.5 PH	+0.01 PH +0.1 PH +0.5 PH	+0.01 PH +0.1 PH +0.5 PH

Notes: H = half-strength Hoagland nutrient solution; $H(-P)$ = not contain phosphorus in H; 0.01 PH = 0.01 mg·L^{-1} phosphorus in H; 0.1 PH = 0.1 mg·L^{-1} phosphorus in the $H(-P)$; 0.5 PH = 0.5 mg·L^{-1} phosphorus in the $H(-P)$.

2.5. Data Analysis and Statistical Analysis

Fe (Mn, P) content (g) = Fe (Mn, P) concentration × dry matter.

Concentration of Fe plaque (µg/g) = Fe plaque content/roots dry matter.

Translocation of Fe, P, Mn from roots to the stems and leaves, and element aquisition efficiency (FeAE, MnAE, PAE) were calculated as follows:

Translocation of Fe (Mn, P) in stems (%) = Fe (Mn, P) content in stems/total Fe (Mn, P) content in plants × 100.

Translocation of Fe (Mn, P) in leaves (%) = Fe (Mn, P) content in leaves/total Fe (Mn, P) content in plants × 100.

Fe acquisition efficiency (FeAE, mg·g^{-1}) = total Fe content in plants/the root dry matter.

Mn acquisition efficiency (Mn AE, mg·g^{-1}) = total Mn content in plants/the root dry matter.

P acquisition efficiency (PAE, mg·g^{-1}) = total P content in plants/the root dry matter.

Statistical analyses were performed with SPSS Statistics 20.0. Two-way ANOVA tests were performed according to any interactions of the Fe and P treatments. The Kolmogorov-Smirnov test was used to test for normality. The data of iron plaque were square root transformed before parametric analysis. The Levene's test was performed to test for the homogeneity of variances. Significant differences between treatments in means were compared using Tukeys's with significance set at $p \leq 0.05$.

3. Result

3.1. Iron Plaque Formation on the Root

DCB-extractable Fe concentrations from the roots of *G. spiculosa* varied from 2.1–233 µg·g^{-1}, with the highest value recorded at the Fe 500/P 0.01 treatment, and the lowest value at Fe 0/P 0.5 treatment (Figure 1). The iron plaque was visually present on the root surface of Fe 20–Fe 500. There was a sharp increase of iron plaque at the concentration of Fe 500, which was about seven times higher than that of Fe 100. In both Fe 0 and Fe 1 treatments, there were no statistically significant differences in the formation of iron plaque. With the increase of Fe^{2+} applied, the amount of iron plaque with P 0.01 were higher than P 0.1 and P 0.5 for 20–500 mg·L^{-1} Fe treatments.

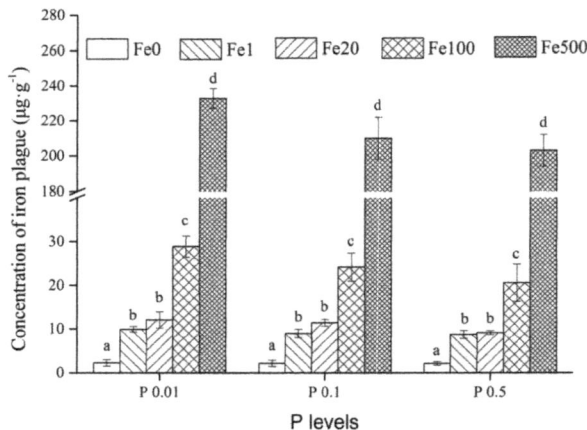

Figure 1. The amount of iron plaque on roots surface of *G. spiculosa* grown hydroponically with varying Fe^{2+} concentrations and P levels. Values (mean ± SE, $n = 3$) followed by different letters designate significant differences ($p < 0.05$) between Fe^{2+} concentrations for a given P level (a, b, c, d). The data were square root transformed before statistical analysis.

3.2. Dry Matter Accumulation

Dry matter of plants were significantly affected by Fe and P concentrations (Table 2). In plants exposed to Fe 20 and above, the dry matter of roots were decreased considerably, whereas the dry matter of stems and leaves showed similar response patterns, peaking at Fe 20 and subsequently decreasing with increasing Fe application, with stems ranging from 0.47 g (Fe 500/P 0.1) to 1.10 g (Fe 20/P 0.01) and leaves ranging from 0.15 g (Fe 500/P 0.1) to 0.93 g (Fe 20/P C.01). The highest plant biomass at the three P levels was mostly observed at P 0.01 and a few appeared at P 0.5 treatment. The results showed that plants have different biomass distribution patterns for organs with varying Fe and P concentrations.

Table 2. Dry matter of *G. spiculosa* exposed to different concentrations of Fe and P (mean ± SE, $n = 5$), and the analysis of variance (two-way ANOVA) on dry matter.

Dry Matter (g)							
Treatment	Root	Stem	Leave	Treatment	Root	Stem	Leave
Fe 0/P 0.01	0.51 ± 0.03 b	0.59 ± 0.02 ab	0.52 ± 0.05 bc	Fe 1/P 0.01	0.54 ± 0.02 b	0.83 ± 0.0≤ c	0.69 ± 0.03 c
Fe 0/P 0.1	0.43 ± 0.04 b	0.64 ± 0.04 b	0.48 ± 0.03 b	Fe 1/P 0.1	0.39 ± 0.05 b	0.83 ± 0.05 c	0.72 ± 0.03 c
Fe 0/P 0.5	0.44 ± 0.03 bc	0.79 ± 0.06 b	0.56 ± 0.02 bc	Fe 1/P 0.5	0.33 ± 0.04 bc	0.86 ± 0.0≤ b	0.67 ± 0.05 cd
Fe 20/P 0.01	0.46 ± 0.03 b	1.10 ± 0.10 d	0.93 ± 0.08 d	Fe 100/P 0.01	0.43 ± 0.04 b	0.79 ± 0.08 bc	0.49 ± 0.02 b
Fe 20/P 0.1	0.34 ± 0.03 b	0.86 ± 0.07 c	0.80 ± 0.09 c	Fe 100/P 0.1	0.28 ± 0.04 b	0.54 ± 0.04 b	0.41 ± 0.05 b
Fe 20/P 0.5	0.46 ± 0.03 c	0.85 ± 0.06 b	0.76 ± 0.04 d	Fe 100/P 0.5	0.35 ± 0.03 b	0.67 ± 0.06 ab	0.42 ± 0.02 b
Fe 500/P 0.01	0.18 ± 0.03 a	0.55 ± 0.02 a	0.27 ± 0.02 a		Root	Stem	Leave
Fe 500/P 0.1	0.09 ± 0.01 a	0.47 ± 0.03 a	0.15 ± 0.02 a	Fe	<0.001	<0.001	<0.001
Fe 500/P 0.5	0.19 ± 0.02 a	0.50 ± 0.06 a	0.22 ± 0.03 a	P	<0.01	<0.05	NS
				Fe × P	NS	<0.01	NS

Note: Different letters designate significant differences ($p < 0.05$) between Fe^{2+} concentrations for a given P level (a, b, c, d).

3.3. SEM and EDS Analysis

The morphological analysis of roots by SEM showed that iron plaque clearly visible as an orange-brown deposition on the root surfaces. Figure 2A showed the root surface of Fe 0 treatment, which was the lowest deposition with plaque. The coating was extensively distributed across the root surface and consistently increased with increasing Fe^{2+} (Figure 2B–E,b–e). Microorganisms were also observed on the root surfaces of *G. spiculosa*, such as Figure 2a,b.

To confirm these visual observations, further examination was conducted by Energy Dispersive X-ray Spectrometry (EDS) scanning. Different colors were added to the image representing various elements. Figure 3 shows bright particles deposited on the root surfaces of Fe 500/P 0.1. In this image, Fe is represented by red, P by green, Mn by purple, and S by blue. The element of Fe was more dense than P and Mn.

3.4. Uptake and Accumulation of Fe, Mn, P

For each tissue type, the patterns of Fe content were similar for the same Fe and P treatments (Figure 4). Increasing Fe concentrations significantly increased the Fe content of roots, stems, and leaves. The highest Fe contents for Fe 0–Fe 20 treatments were found in the roots with 1.10–3.97 mg (stems with 0.33–2.11 mg and leaves with 0.10–0.46 mg); however, Fe 100 and Fe 500 were found in the stems with 9.63–27.87 mg (roots with 5.92–15.22 mg and leaves with 1.13–4.95 mg). At the three P levels in a given Fe concentration, the total Fe content of P 0.01 in Fe 0–Fe 20 treatment was higher than the other P levels, and higher in the Fe 100/P 0.5 and Fe 500/P 0.1 treatments.

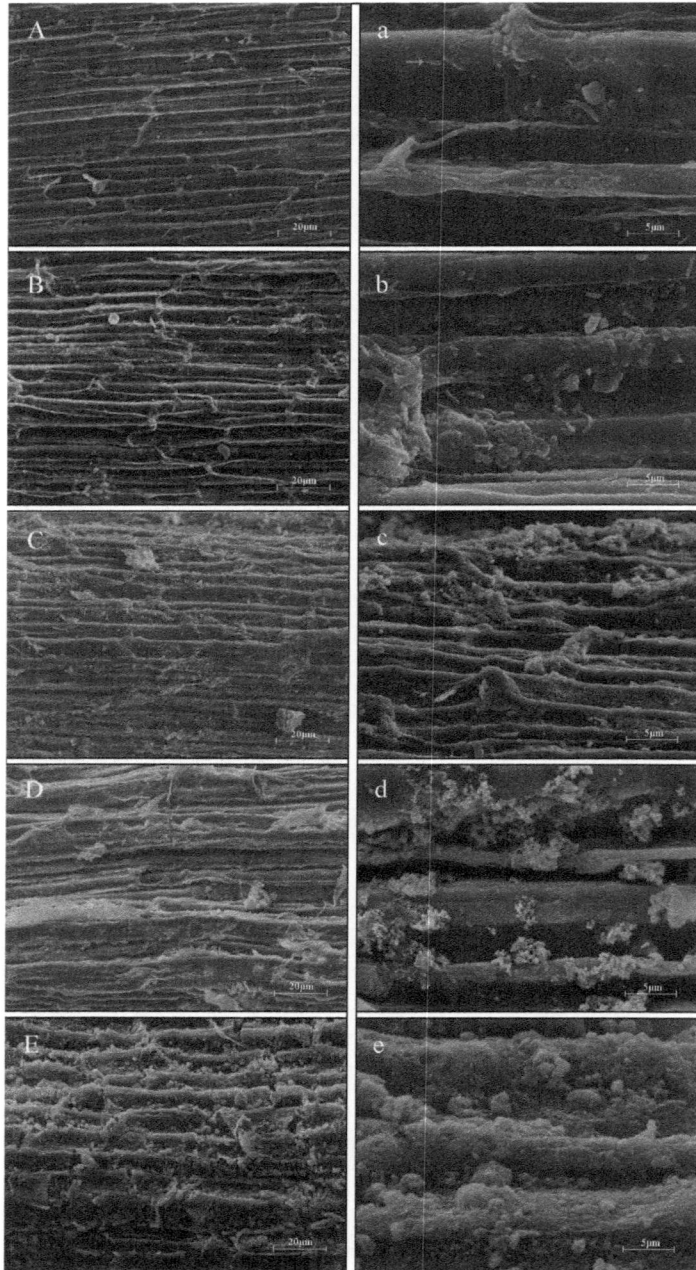

Figure 2. Different amounts of iron plaque formed on the roots of *G. spiculosa* observed by Scanning Electron Microscopy (SEM), accuracy of 20 μm: (**A–E**), accuracy of 40 μm: (**a–e**). Five treatments were represented: Fe 0/P 0.1, Fe 1/P 0.1, Fe 20/P 0.1, Fe 100/P 0.1, Fe 500/P 0.1. Red arrows indicated microorganisms exsiting on the roots surface.

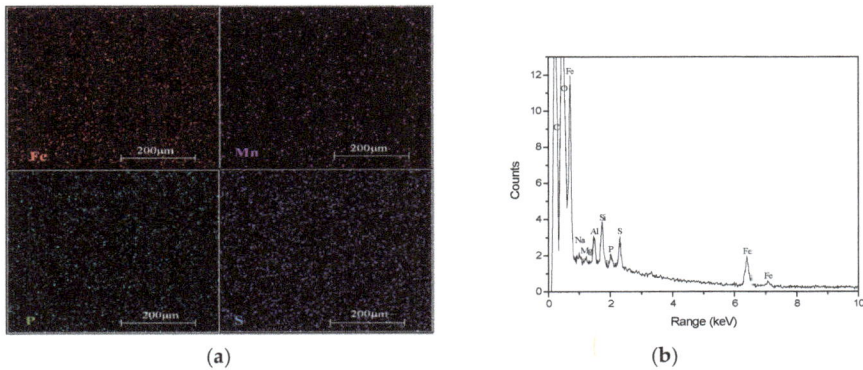

(a)   (b)

Figure 3. Energy Dispersive X-ray Spectrometry (EDS) on the root surface of *G. spiculosa* distribution of Fe, Mn, P, and S (**a**), and a quantitative determination of elements composition (**b**).

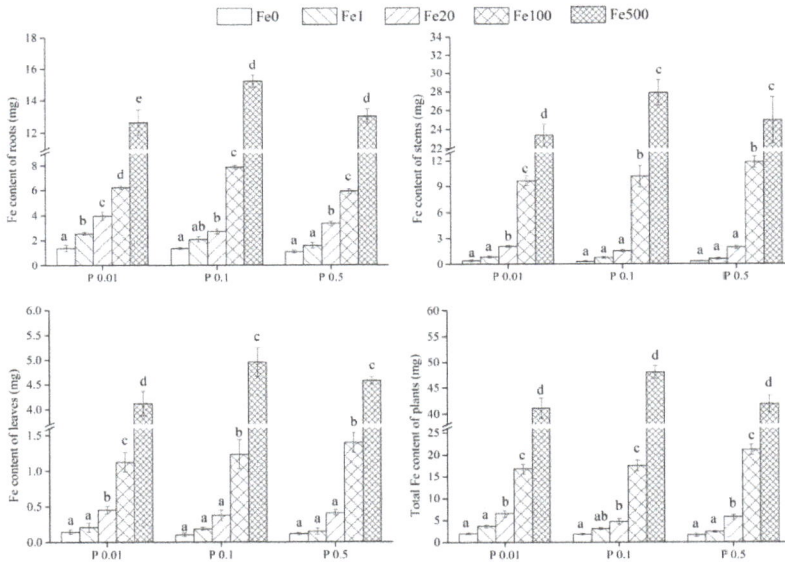

Figure 4. Effects of Fe^{2+} and P on roots, stems, leaves, and total Fe contents of *G. spiculosa*. Values (mean ± SE, *n* = 3) followed by different letters designate significant differences ($p < 0.05$) between Fe^{2+} concentrations for a given P level (a, b, c, d).

Mn content in stems was significantly higher than in roots (1.04–4.35 times) and leaves (1.03–2.31 times) (Figure 5). In contrast to Fe and P, total Mn content of Fe 0 and Fe 1 treatments at any given P level was greater than others. The total Mn content in P 0.01 of Fe 0–Fe 20 was higher than P 0.1 and P 0.5 treatments, and higher in the Fe 100/P 0.5 and Fe 500/P 0.1 treatments. Plants grown in the Fe 500 treatment had the lowest total Mn content (23.1%–41.1% of other treatments).

Stems (1.6–3.7 times) and leaves (1.4–2.9 times) had greater P content than roots for the same treatments (Figure 6). P content in each tissue type increased with Fe^{2+} addition until Fe 100 and Fe 500 treatments, which showed a downward trend. The total P content of P 0.01 treatment was higher than the other P levels except the Fe 0 treatment. Among five Fe concentrations at a given P level, total P content of Fe 1 was greater and greatest in the Fe 20 treatments.

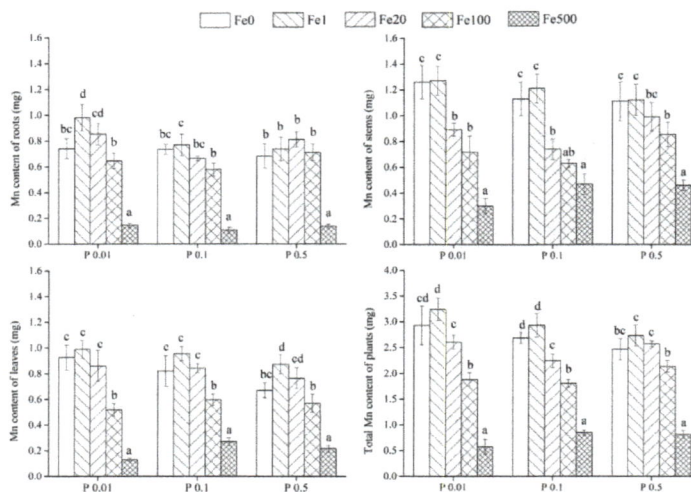

Figure 5. Effects of Fe^{2+} and P on roots, stems, leaves, and total Mn contents of *G. spiculosa*. Values (mean ± SE, *n* = 3) followed by different letters designate significant differences (*p* < 0.05) between Fe^{2+} concentrations for a given P level (a, b, c, d).

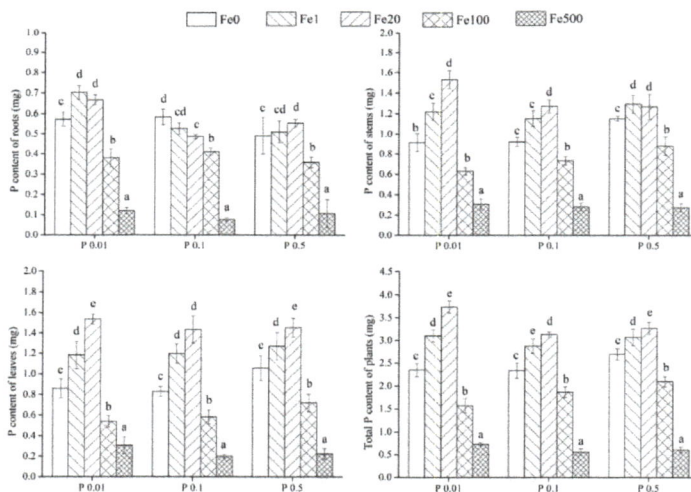

Figure 6. Effects of Fe^{2+} and P on roots, stems, leaves and total P contents of *G. spiculosa*. Values (mean ± SE, *n* = 3) followed by different letters designate significant differences (*p* < 0.05) between Fe^{2+} concentrations for a given P level (a, b, c, d).

There were close correlations between iron plaque and total Fe, Mn, P content in plants. For total Fe content, a positive correlation was observed with iron plaque (Figure 7a, r^2 = 0.86). Curviliner correlation between iron plaque and total Mn content (Figure 7c, r^2 = 0.89) was significant. The total P content in plants were also significantly correlated with iron plaque (Figure 7b, r^2 = 0.84).

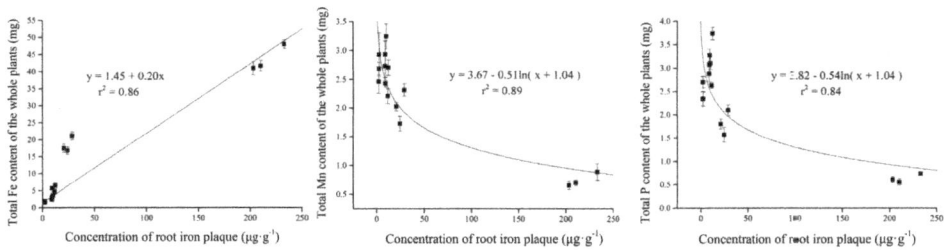

Figure 7. Relationship between the concentration of root iron plaque and total Fe content, total P content, and total Mn content of the whole plant. Data were means ± SE (*n* = 3).

3.5. Translocation and Acquisition Efficiency of Fe, Mn, and P

As expected, Fe translocation in stems was significantly increased with the increasing iron in the solution (Table 3). The distribution of Fe in leaves was obviously lower than stems. The P translocation in stems and leaves was not as significant as in Fe, but data of Fe 1 and Fe 20 concentrations were higher than other Fe treatments. The Mn translocation in stems and leaves was similar between those treatments, and the distribution of Mn (22.4%–37.5%) and P (30.0%–44.9%) in leaves was higher than that of Fe (5.7%–10.9%).

The responses of FeAE were similar to the Fe translocation between Fe treatments for a given P level, which showed a positive correlation with Fe applied. For the difference of FeAE between three P levels, higher values were found in P 0.1 treatment of Fe 0–Fe 500, except there was no difference found in Fe 20. The results of MnAE and PAE were similar in terms of translocation, with higher MnAE in Fe 1 and PAE in Fe 1–Fe 20 treatments. For three P levels, the difference between FeAE, MnAE and PAE were not significant.

Table 3. Effect of Fe and P on the translocation of Fe, Mn, and P from roots to the stems and leaves, and acquisition efficiency of Fe, Mn, and P (FeAE, MnAE, and PAE).

Treatements		Translocation (%) and Acquisition Efficiency of Fe, Mn, and P (FeAE, MnAE, and PAE, mg·g⁻¹)								
		Fe			Mn			P		
Fe	P	Stems	Leaves	FeAE	Stems	Leaves	MnAE	Stems	Leaves	PAE
Fe 0	P 0.01	20.5 ± 3.3 a	7.6 ± 1.6 ab	3.8 ± 0.3 a	43.0 ± 6.7 a	31.6 ± 0.7 a	5.8 ± 0.2 bc	38.9 ± 7.9 a	36.6 ± 9.2 a	4.6 ± 0.3 ab
	P 0.1	18.5 ± 4.9 a	5.7 ± 0.8 a	4.1 ± 0.2 a	42.1 ± 9.4 ab	30.5 ± 0.9 ab	6.3 ± 0.4 ab	39.5 ± 4.4 a	35.5 ± 4.6 a	5.5 ± 0.5 ab
	P 0.5	21.8 ± 2.5 a	7.5 ± 0.7 a	3.7 ± 0.4 a	45.1 ± 6.7 a	27.2 ± 0.5 a	5.6 ± 0.4 b	42.6 ± 2.2 a	39.2 ± 9.2 a	6.2 ± 0.5 bc
Fe 1	P 0.01	23.9 ± 6.4 a	5.8 ± 0.7 a	6.8 ± 0.2 ab	39.2 ± 6.6 a	30.5 ± 1.2 a	7.6 ± 0.3 c	39.2 ± 8.3 a	38.1 ± 10.2 a	8.8 ± 0.2 b
	P 0.1	25.4 ± 7.2 a	6.1 ± 0.5 ab	7.9 ± 0.6 b	41.2 ± 8.0 ab	32.6 ± 3.5 ab	7.4 ± 0.8 b	40.1 ± 7.7 a	41.6 ± 9.9 ab	7.4 ± 0.9 bc
	P 0.5	26.6 ± 8.1 a	6.3 ± 0.7 a	7.2 ± 0.4 b	41.0 ± 5.1 a	32.0 ± 6.7 a	7.3 ± 0.7 c	42.0 ± 8.7 a	41.4 ± 4.7 a	8.3 ± 0.7 d
Fe 20	P 0.01	32.2 ± 6.3 a	7.0 ± 0.9 a	14.0 ± 0.7 b	34.2 ± 3.8 a	33.1 ± 5.1 a	5.9 ± 0.2 bc	41.0 ± 9.1 a	41.2 ± 4.8 a	8.1 ± 0.8 c
	P 0.1	33.6 ± 5.6 a	8.1 ± 1.1 bc	13.8 ± 0.8 c	32.9 ± 5.6 a	37.5 ± 6.5 b	6.6 ± 0.3 ab	39.8 ± 6.3 a	44.9 ± 6.9 b	7.8 ± 0.6 c
	P 0.5	34.2 ± 8.4 a	7.2 ± 0.7 a	12.5 ± 0.5 c	38.5 ± 3.5 a	29.8 ± 4.5 a	5.3 ± 0.2 b	38.7 ± 9.4 a	44.4 ± 10.5 a	7.1 ± 0.5 c
Fe 100	P 0.01	57.4 ± 9.3 b	6.7 ± 0.4 a	49.0 ± 2 c	38.3 ± 9.5 a	27.7 ± 0.9 a	5.4 ± 0.5 b	44.5 ± 4.3 a	34.5 ± 9.3 a	4.9 ± 0.3 ab
	P 0.1	58.1 ± 7.2 b	7.0 ± 0.6 ab	61.2 ± 4.7 d	34.8 ± 3.0 a	33.1 ± 1.1 ab	6.3 ± 0.6 ab	40.1 ± 3.6 a	37.2 ± 5.7 a	5.7 ± 0.4 ab
	P 0.5	56.1 ± 6.5 b	6.6 ± 0.8 a	51.8 ± 1.8 d	39.9 ± 4.8 a	26.8 ± 3.8 a	5.8 ± 0.5 b	48.9 ± 9.3 a	30.0 ± 6.8 a	5.1 ± 0.2 b
Fe 500	P 0.01	56.9 ± 8.2 b	10.0 ± 1.3 b	265.7 ± 44.1 d	51.7 ± 4.0 a	22.4 ± 5.4 a	4.8 ± 0.1 a	41.8 ± 5.5 a	41.8 ± 8.7 a	3.9 ± 0.6 a
	P 0.1	58.0 ± 7.9 b	10.3 ± 0.8 c	350.4 ± 18.9 e	55.3 ± 6.2 b	31.8 ± 0.4 a	5.3 ± 0.3 a	50.3 ± 3.8 a	35.9 ± 1.8 a	4.6 ± 0.3 a
	P 0.5	59.5 ± 5.7 b	10.9 ± 1.6 a	280.3 ± 16.6 e	56.1 ± 6.3 a	26.4 ± 2.5 a	3.5 ± 0.5 a	45.1 ± 4.1 a	37.4 ± 5.2 a	3.3 ± 0.2 a

Notes: The data of FeAE were square transformed before parametric analysis. Data are means ± SE ($n = 3$). For each parameter values followed by different letters designate significant differences ($p < 0.05$) between Fe^{2+} concentrations for a given P level (a, b, c, d).

4. Discussion

4.1. Iron Plaque Formation, SEM and EDS Analysis

Iron in the aquatic environment can not only directly participates in the physiological processes of wetland plants, but also forms iron plaque indirectly affecting the absorption and translocation of nutrients. Some research has shown that nutrient limitation may be the major factor to the growth of wetland plants [25]. In present study, the physico-chemical properties of iron plaque of *G. spiculosa* were characterized by SEM-EDX (Figures 2 and 3). The elements present in the plaque included essential plant nutrients, such as Fe, O, Mn, P, S, Mg, and Al. The amounts of iron plaque on the root surface were significantly affected by amounts of Fe supplied in the culture medium. Iron plaque formation on the roots increased with increasing levels of iron in the growth medium. Additionally, the root dry matter of *G. spiculosa* was declining with the increasing iron additon (Table 2), indicating that large amount of iron plaque harmed the growth of plant roots. Our results further showed that iron plaque formation was highest at the lower P treatments, which is comparable with data reported for rice [9,26]. Fu et al. [27] reported that P deficiency in the rhizosphere could increase the oxidizing capability of rice roots, which was associated with production of reactive oxygen species, increase in antioxidant enzyme activity, and the release of O_2 and oxidative substances from the root. The Fe^{2+} around the rhizosphere is oxidized by these oxidizing substances to Fe^{3+} that precipitates as orange iron plaque accumulation on the root surface of aquatic plants and on soil particles in the rhizosphere [28]. For wetland species, however, few studies focus on the iron plaque formation on the root surface under P-deficiency conditions [9,26].

4.2. Uptake, Translocation of Fe and FeAE

In wetland environments, Fe immobility is highly variable. Plants therefore need a tight regulation of Fe uptake, transport, and storage to ensure balanced development by avoiding both Fe deficiency and toxicity [29]. The total Fe contents of the plants significantly increased with the application of Fe. Meanwhile, our results indicated that wetland plants have high innate tolerance to iron in their habitats, which demonstrates the reasons for growing wetland plants in iron ore mining and other iron-contamined environments [30]. All plants were collected from areas that had not been exposed to high levels of Fe^{2+}. As a result, Fe tolerance would have to be innate. In response to Fe overload, plants express a series of Fe homeostasis-related genes, including those encoding for ferritin and a common protein for Fe storage [31]. The transporters are generally divided into low-affinity Fe transport systems and high-affinity Fe transport systems. Low-affinity transport systems take up sufficient Fe in high levels of Fe, meanwhile preventing Fe overload. In contrast, high-affinity Fe transport systems are activated by Fe deficiency conditions [32,33]. The formation of iron plaque under P-deficiency conditions had a large effect on Fe uptake of roots. Furthermore, the expression of ferritin genes in *Arabidopsis* plants can be effectively induced by P-deficiency conditions, such as encoding of NAS3 and ZIP5 for metal transport [21]. Moreover, the induction of genes involved in the transport and storage inside the plant in the forms of chelated iron, by metallothioneins, nicotianamine, and ferritin, suggesting a global and consistent plant response to P-deficiency [34].

Our results futher suggested the translocation was affected. The Fe translocation process involves various Fe transport components, such as those involved with long-distance transport within the vascular system (from root to shoot, cotyledons to young shoots), and short-distance transport within the cells (intracellularly in the symplast), to final destination targets within cells (compartment or biomolecules) [29]. These transport processes and strategies aim to protect and maintain plants' health. In this study, with increasing Fe supply, the content and translocation of Fe in the stems gradually increased, but Fe in leaves remained stable (5.7%–8.1%), except for the highest exposure (Fe 500: 10.0%–10.9%). Wetland plants seem to generally engage in an "exclusion" strategy, in which metals in aboveground parts are maintained at a constant low levels until a critical external concentration is reached [10]. In our experiment, leaf-wilting and desiccation were observed at the Fe 500 treatment, and those are well-known symptoms, which may have caused unrestricted metal transport. Excessive absorption of Fe^{2+} by roots and subsequent translocation to leaves causes an elevated production of toxic oxygen radicals, which destroys cell structural components and damages physiological processes [35]. The critical Fe content depends on plant age. The color of leaves could be changed from orange to rusty brown with excessive Fe content in plant, especially in older leaves [36,37]. It is most likely that the result of damage to the root system, and the rapid reduction of root biomass was the best evidence. Consistent with our study, Batty et al. [38] found that the roots of aquatic plants were inhibited under conditions of excessive Fe supply, including root flaccidity and reduced root branching.

4.3. Uptake, Translocation of Mn and MnAE

Mn is essential for plant metabolism and develpment, and exists in oxidation states II, III, and IV inside about 35 enzymes of a plant cell [39]. Mn mainly achieves two functions in proteins: (1) it acts as a catalyzing active metal, or (2) it shows an activating role on enzymes [40]. The concentration of Mn in the nutrient solution was suitable for plant growth, so it can be used to detect the effects of Fe and P on the absorption of micronutrients. The Mn content was increased with the treatment of $1 \text{ mg} \cdot L^{-1}$ Fe, which concentration was close to the natural habitat of the plants. The results indicated that large amount of agricultural drainage water containing iron-rich groundwater would promote the growth of wetland plants, due to the high iron tolerance of plants. The results confirmed that the Mn uptake was promoted by iron addition within the natural range, but significantly suppressed when iron concentration become excessively high. The mechanism by which plaque affects the precipitation and uptake of Mn, Pb, and Zn may involve the absorption of metals onto the large surface plaque area [12,41]. Inside the plants, an Fe transport protein (RcITP) binds preferentially Fe(III) over Fe(II), but it also forms compleses with Mn, Zn, and Cu, which also promote Mn absorption and migration at the same time [42]. Kruger et al. [42] further showed that a thick iron plaque may have limited capacity for the absorption of Mn and other metals, which was consistent with our study. Other results have also shown that the concentrations of Mn tends to be reduced under similar conditions, suggesting that uptake is particularly affected by large amounts of iron plaque [43]. Batty et al. [41] found that the Mn concentration in the absence of plaque was higher than when plaque was present. The EDX analysis in our study also showed that large amounts of Mn existed on the surface of the plaque, suggesting that abundant iron plaque acted as a barrier for micronutrients by forming a complex precipitation or adsorption on the plaque surface.

The absorption and translocation of micronutrients seemed to be sensitive to high concentrations of Fe application. Studies with Lynch and Clair [44], and Socha and Guerinot [45], have shown that Mn deficiency can be caused by multiple factors including high concentrations of other minerals in the soil, such as Fe, Mg, Ca, and P. The results of Ghasemi-Fasaei et al. [46] showed that foliar application of Fe on chickpea decreased Mn uptake owing to the antagonistic effect of Fe on translocation of Mn from root to shoot. Similar to Mn, the uptake and translocation of Zn was also inhibited by Fe in the solution [47,48]. Mn, Zn, or Cu deficiency may be induced following excessive Fe fertilization due to uptake of relatively large amounts of Fe [49]. Hayes et al. [48], showed a negative relationship between leaf Mn and total soil P concentration along a bay in western Australia. Soluble P in the rhizosphere can form coprecipitation with Fe^{3+} and Mn^{4+}, which can bind to amorphous oxyhydroxides in root plaques. The conclusion was consistent with our results: total Mn content in plants of P 0.5 was lower than the other P treatments. In general, plants receive higher MnAE by roots and transportation to aboveground with the Fe 1 treatment.

4.4. Uptake, Translocation of P and PAE

Excessive fertilization of agricultural land has resulted in large losses of P from paddy field to the aquatic environment. In recent years, interactions between Fe and P have already been utilized in technological applications. For example, studies have shown that Fe-coated sand as a liner in drainage systems could reduce P loading of surface water [50]. The formation of iron plaque on the root surface of wetlands has a number of functions, such as nutrient and contaminant uptake. Khan et al. [5] reported that iron plaque has a high capacity to bind P in wetlands, and the sediments Eh, P, and Fe cycling have close links in lake ecosystems. Furthermore, Fe sludge and Fe-rich groundwater could be used as reactive materials to control the eutrophication water bodies, which could bound a large amount of P [23]. Apart from the potential role of iron plaque in changing the bioavailability of metals in the rhizosphere of wetland plants, iron plaque is believed to be important in plant P uptake and translocation. In our experiment, P 0.01 during the last two weeks was intended to simulate the P-deficient condition of *G. spiculosa* grown in solution. However, in retrospect, the plants may have obtained sufficient amounts of P from the nutrient solution in the initial three weeks, also aided by P adsorbed in the iron plaque. Similar results were found for common reed (*Phragmites australis* (Cav.) Trin ex. Steudel.), in which the amount of P adsorbing increased with the amount of plaque formation on the roots [8,38]. Even without P supplied in the solution, P concentrations in rice seedings were not deficient at the background levels of Fe applied [9].

Hu et al. [22] showed that P deficiency in habitats results in an increase in plant root length and porosity, thereby increasing aquatic plants ROL that subsequently would oxidize higher amounts of Fe^{2+} and form root plaque. Although previous studies indicated that root porosity, ROL, and Fe plaque are mainly controlled genetically [51], wastewater as an environmental stress or can change these properties. High strength wastewater, such as the agricultural drainage water from paddy fields, significantly decreased ROL and root porosity, but high level of iron promoted the formation of iron plaque, further enhancing the absorption of P from habitats. The success of iron addition in order to regulate P release can be calculated by using the Fe:P ratio in the sediment pore water, which ranging from 1–15 of pore water ratio [52,53]. The pore water Fe:P ratio provides a good prognosis for both the restoration of water quality and biodiversity after restoration measures. Therefore, these reactive materials can be used for trapping P in the soil solution by using iron rich groundwater for irrigation or adding the reactive materials in lakes or wetlands when agricultural drainage water exports into the surrounding ecosystems.

Apart from enhacing P absorption from habitats, iron plaque promoted P translocation to the aboveground. Unlike other elements, the P content of stems and leaves was nearly two times that of the roots for a given Fe and P treatment (Figure 6), and the PAE of Fe 1–Fe 20 showed much higher values. P plays a crucial role in plant cell division, growth, and biomass accumulation. Our results implied that iron-rich groundwater flowing into wetlands may be beneficial to P uptake and transfer to aboveground biomass in wetland plants. However, the P content was reduced significantly when the external Fe concentration was over $100 \ mg \cdot L^{-1}$ in our study, partly because the roots of plants were harmed and large amounts of iron oxyhydroxides in the root plaque bound sequestrated P. Stimulation of development of iron plaque on the root surface has been used to reduce the toxic effects of As, Pb, and Cd, and the effect of P decreased metal desorption, solubility, and bioavailability in the rhizosphere; this even enhanced tolerance to heavy metals [20,54].

5. Conclusions

Our findings provided mechanistic insights into Fe biogeochemistry in the rhizosphere of *G. spiculosa* in relation to P uptake, Mn, and other micronutrients. The presence of excess Fe^{2+} in the aquatic environment induced the formation of iron plaque on the root surface. The amount of iron plaque increased with increasing Fe^{2+} supply, and P-deficiency promoted its production. Iron plaque on the root surface could function as a reservoir for nutrients, but inhibited element accumulation and translocation in the plants when present as a thick cover on the roots. In the presence of excessive iron plaque, the uptake and translocation of Fe in plants was decreased significantly. The absorption of Mn was particularly affected by iron plaque. Formation of iron plaque enhanced the P utilization until the external iron concentration exceeded $100 \ mg \cdot L^{-1}$.

Acknowledgments: We sincerely thank Beth A Middleton of United States Geological Survey for her constructive comments on the early version of this manuscript. This research was supported by Minstry of Science and Technology of People's Republic of China (2016YFA0602303), the National Key Research & Development Program of China (2016YFC0500408), National Natural Science Foundation of China (41771120, 41271107, 41471079), and the Northeast Institute of Geography and Agroecology, CAS (IGA-135-05).

Author Contributions: Conceived and designed the experiments: X.J., M.J., Y.Z., and X.L. Performed the experiments: X.J., Y.L., X.T., L.Q. Analyzed the data and prepared the manuscript: X.J., M.J. and M.L.O. revised it. All authors read and approved the manuscript.

Conflicts of Interest: All authors have no conflict of interests to any other parties.

References

1. Messer, T.L.; Burchell, M.R., II; Birgand, F.; Broome, S.W.; Chescheir, G. Nitrate removal potential of restored wetlands loaded with agricultural drainage water: A mesocosm scale experimental approach. *Ecol. Eng.* **2017**, *106*, 541–554. [CrossRef]
2. Dunne, E.J.; Coveney, M.F.; Hoge, V.R.; Conrow, R.; Naleway, R.; Lowe, E.F.; Battoe, L.E.; Wang, Y. Phosphorus removal performance of a large-scale constructed treatment wetland receiving eutrophic lake water. *Ecol. Eng.* **2015**, *79*, 132–142. [CrossRef]
3. Vymazal, J.; Brezinova, T. The use of constructed wetlands for removal of pesticides from agricultural runoff and drainage: A review. *Environ. Int.* **2015**, *75*, 11–20. [CrossRef] [PubMed]
4. Wang, C.Y.; Sample, D.J.; Day, S.D.; Grizzard, T.J. Floating treatment wetland nutrient removal through vegetation harvest and observations from a field study. *Ecol. Eng.* **2015**, *78*, 15–26. [CrossRef]
5. Khan, N.; Seshadri, B.; Bolan, N.; Saint, C.P.; Kirkham, M.B.; Chowdhury, S.; Yamaguchi, N.; Lee, D.Y.; Li, G.; Kunhikrishnan, A.; et al. Chapter one-root iron plaque on wetland plants as a dynamic pool of nutrients and contaminants. *Adv. Agron.* **2016**, *138*, 1–96.
6. Allen, W.C.; Hook, P.B.; Biederman, J.A.; Stein, O.R. Temperature and wetland plant species effects on wastewater treatment and root zone oxidation. *J. Environ. Qual.* **2002**, *31*, 1010–1016. [CrossRef] [PubMed]
7. Liang, Y.; Zhu, Y.G.; Xia, Y.; Li, Z.; Ma, Y. Iron plaque enhances phosphorus uptake by rice (*Oryza sativa*) growing under varying phosphorus and iron concentrations. *Ann. Appl. Biol.* **2006**, *149*, 305–312. [CrossRef]
8. Liu, W.J.; Zhu, Y.G.; Smith, F.A.; Smith, S.E. Do phosphorus nutrition and iron plaque alter arsenate (As) uptake by rice seedlings in hydroponic culture? *New Phytol.* **2004**, *162*, 481–488. [CrossRef]
9. Kobayashi, T.; Nishizawa, N.K. Iron uptake, translocation, and regulation in higher plants. *Annu. Rev. Plant. Biol.* **2012**, *63*, 131–152. [CrossRef] [PubMed]
10. Sebastian, A.; Prasad, M.N. Iron plaque decreases cadmium accumulation in *Oryza sativa* L. and serves as a source of iron. *Plant Biol.* **2016**, *18*, 1008–1015. [CrossRef] [PubMed]
11. Hansel, C.M.; Fendorf, S.; Sutton, S.; Newville, M. Characterization of Fe plaque and associated metals on the roots of mine-waste impacted aquatic plants. *Environ. Sci. Technol.* **2001**, *35*, 3863–3868. [CrossRef] [PubMed]
12. Machado, W.; Gueiros, B.B.; Lisboa-Filho, S.D.; Lacerda, L.D. Trace metals in mangrove seedlings: Role of iron plaque formation. *Wetlands Ecol. Manag.* **2005**, *13*, 199–206. [CrossRef]
13. Xu, D.; Xu, J.; He, Y.; Huang, P.M. Effect of iron plaque formation on phosphorus accumulation and availability in the rhizosphere of wetland plants. *Water Air Soil. Pollut.* **2008**, *200*, 79–87. [CrossRef]
14. Ye, Z.; Baker, A.J.; Wong, M.H.; Willis, A.J. Zinc, lead and cadmium accumulation and tolerance in *Typha latifolia* as affected by iron plaque on the root surface. *Aquat. Bot.* **1998**, *61*, 55–67. [CrossRef]
15. Tripathi, R.D.; Tripathi, P.; Dwivedi, S.; Kumar, A.; Mishra, A.; Chauhan, P.S.; Norton, G.J.; Nautiyal, C.S. Roles for root iron plaque in sequestration and uptake of heavy metals and metalloids in aquatic and wetland plants. *Metallomics* **2014**, *6*, 1789–1800. [CrossRef] [PubMed]
16. Otte, M.; Rozema, J.; Koster, L.; Haarsma, M.; Broekman, R. Iron plaque on roots of *Aster tripolium* L.: Interaction with zinc uptake. *New Phytol.* **1989**, *111*, 309–317. [CrossRef]
17. Zhang, X.; Zhang, F.; Mao, D. Effect of iron plaque outside roots on nutrient uptake by rice (*Oryza sativa* L.): Phosphorus uptake. *Plant Soil* **1999**, *209*, 187–192. [CrossRef]
18. Raghothama, K. Phosphate acquisition. *Annu. Rev. Plant Biol.* **1999**, *50*, 665–693. [CrossRef] [PubMed]
19. Lambers, H.; Hayes, P.E.; Laliberte, E.; Oliveira, R.S.; Turner, B.L. Leaf manganese accumulation and phosphorus-acquisition efficiency. *Trends Plant Sci.* **2015**, *20*, 83–90. [CrossRef] [PubMed]

20. Yang, Y.; Chen, R.; Fu, G.; Xiong, J.; Tao, L. Phosphate deprivation decreases cadmium (Cd) uptake but enhances sensitivity to Cd by increasing iron (Fe) uptake and inhibiting phytochelatins synthesis in rice (*Oryza sativa*). *Acta Physiol. Plant.* **2015**, *38*, 28. [CrossRef]

21. Hirsch, J.; Marin, E.; Floriani, M.; Chiarenza, S.; Richaud, P.; Nussaume, L.; Thibaud, M.C. Phosphate deficiency promotes modification of iron distribution in *Arabidopsis* plants. *Biochimie* **2006**, *88*, 1767–1771. [CrossRef] [PubMed]

22. Hu, Y.; Huang, Y.Z.; Liu, Y.Z. Influence of iron plaque on chromium accumulation and translocation in three rice (*Oryza sativa* L.) cultivars grown in solution culture. *Chem. Ecol.* **2014**, *30*, 29–38. [CrossRef]

23. Chardon, W.J.; Groenenberg, J.E.; Temminghoff, E.J.; Koopmans, G.F. Use of reactive materials to bind phosphorus. *J. Environ. Qual.* **2012**, *41*, 636–646. [CrossRef] [PubMed]

24. Taylor, G.J.; Crowder, A. Use of the DCB technique for extraction of hydrous iron oxides from roots of wetland plants. *Am. J. Bot.* **1983**, *70*, 1254–1257. [CrossRef]

25. Yamauchi, M. Rice bronzing in Nigeria caused by nutrient imbalances and its control by potassium sulfate application. *Plant Soil* **1989**, *117*, 275–286. [CrossRef]

26. Hu, Y.; Li, J.H.; Zhu, Y.G.; Huang, Y.Z.; Hu, H.Q.; Christie, P. Sequestration of As by iron plaque on the roots of three rice (*Oryza sativa* L.) cultivars in a low-P soil with or without P fertilizer. *Environ. Geochem. Hlth.* **2005**, *27*, 169–176. [CrossRef] [PubMed]

27. Fu, Y.Q.; Yang, X.J.; Shen, H. The physiological mechanism of enhanced oxidizing capacity of rice (*Oryza sativa* L.) roots induced by phosphorus deficiency. *Acta Physiol. Plant.* **2014**, *36*, 179–190. [CrossRef]

28. Yang, J.; Tam, N.F.Y.; Ye, Z. Root porosity, radial oxygen loss and iron plaque on roots of wetland plants in relation to zinc tolerance and accumulation. *Plant Soil* **2013**, *374*, 815–828. [CrossRef]

29. Barton, L.L.; Abadia, J. *Iron Nutrition in Plants and Rhizospheric Microorganisms*; Springer Science & Business Media: Dordrecht, The Netherlands, 2006; pp. 341–357.

30. Matthews, D.J.; Moran, B.M.; Otte, M.L. Screening the wetland plant species *Alisma plantago-aquatica*, *Carex rostrata* and *Phalaris arundinacea* for innate tolerance to zinc and comparison with *Eriophorum angustifolium* and Festuca rubra Merlin. *Environ. Pollut.* **2005**, *134*, 343–351. [CrossRef] [PubMed]

31. Petit, J.M.; Briat, J.F.; Lobréaux, S. Structure and differential expression of the four members of the Arabidopsis thaliana ferritin gene family. *Biochem. J.* **2001**, *359*, 575–582. [CrossRef] [PubMed]

32. Kaplan, J. Strategy and tactics in the evolution of iron acquisition. *Semin. Hematol.* **2002**, *39*, 219–226. [CrossRef] [PubMed]

33. Van Ho, A.; Ward, D.M.; Kaplan, J. Transition metal transport in yeast. *Annu. Rev. Microbiol.* **2002**, *56*, 237–261. [CrossRef] [PubMed]

34. Abel, S. Phosphate sensing in root development. *Curr. Opin. Plant Biol.* **2011**, *14*, 303–309. [CrossRef] [PubMed]

35. Becker, M.; Asch, F. Iron toxicity in rice—Conditions and management concepts. *J. Plant Nutr. Soil Sci.* **2005**, *168*, 558–573. [CrossRef]

36. Yamanouchi, M.; Yoshida, S. *Physiological Mechanism of Rice's Tolerance for Iron Toxicity*; Plant Physiology Departemen of IRRI: Makati, Philippines, 1981; p. 21.

37. Fairhurst, T.; Witt, C.; Buresh, R.; Dobermann, A.; Fairhurst, T. *Rice: A Practical Guide to Nutrient Management*; The International Rice Research Institute (IRRI): Singapore, 2007; pp. 73–75.

38. Batty, L.C.; Younger, P.L. Effects of external iron concentration upon seedling growth and uptake of Fe and phosphate by the common reed, *Phragmites australis* (Cav.) Trin ex. Steudel. *Ann. Bot.-Lond.* **2003**, *92*, 801–806. [CrossRef] [PubMed]

39. Hebbern, C.A.; Laursen, K.H.; Ladegaard, A.H.; Schmidt, S.B.; Pedas, P.; Bruhn, D.; Schjoerring, J.K.; Wulfsohn, D.; Husted, S. Latent manganese deficiency increases transpiration in barley (*Hordeum vulgare*). *Physiol. Plant.* **2008**, *135*, 307–316. [CrossRef] [PubMed]

40. Hansch, R.; Mendel, R.R. Physiological functions of mineral micronutrients (Cu, Zn, Mn, Fe, Ni, Mo, B, Cl). *Curr. Opin. Plant Biol.* **2009**, *12*, 259–266. [CrossRef] [PubMed]

41. Batty, L.; Baker, A.; Wheeler, B.; Curtis, C. The effect of pH and plaque on the uptake of Cu and Mn in *Phragmites australis* (Cav.) Trin ex. Steudel. *Ann. Bot.* **2000**, *86*, 647–653. [CrossRef]

42. Kruger, C.; Berkowitz, O.; Stephan, U.W.; Hell, R. A metal-binding member of the late embryogenesis abundant protein family transports iron in the phloem of *Ricinus communis* L. *J. Biol. Chem.* **2002**, *277*, 25062–25069. [CrossRef] [PubMed]

43. Snowden, R.; Wheeler, B. Chemical changes in selected wetland plant species with increasing Fe supply, with specific reference to root precipitates and Fe tolerance. *New Phytol.* **1995**, *131*, 503–520. [CrossRef]

44. Lynch, J.P.; Clair, S.B.S. Mineral stress: The missing link in understanding how global climate change will affect plants in real world soils. *Field Crop Res.* **2004**, *90*, 101–115. [CrossRef]

45. Socha, A.L.; Guerinot, M.L. Mn-euvering manganese: The role of transporter gene family members in manganese uptake and mobilization in plants. *Front. Plant. Sci.* **2014**, *5*, 106. [CrossRef] [PubMed]

46. Ghasemi-Fasaei, R.; Ronaghi, A.; Maftoun, M.; Karimian, N.A.; Soltanpour, P.N. Iron-manganese interaction in chickpea as affected by foliar and soil application of iron in a calcareous soil. *Commun. Soil Sci. Plan.* **2005**, *36*, 1717–1725. [CrossRef]

47. Martínez-Cuenca, M.R.; Quiñones, A.; Iglesias, D.J.; Forner-Giner, M.Á.; Primo-Millo, E.; Legaz, F. Effects of high levels of zinc and manganese ions on Strategy I responses to iron deficiency in citrus. *Plant Soil* **2013**, *373*, 943–953. [CrossRef]

48. Hayes, P.; Turner, B.L.; Lambers, H.; Laliberté, E. Foliar nutrient concentrations and resorption efficiency in plants of contrasting nutrient-acquisition strategies along a 2-million-year dune chronosequence. *J. Ecol.* **2014**, *102*, 396–410. [CrossRef]

49. Ronaghi, A.; Ghasemi-Fasaei, R. Field evaluations of yield, iron-manganese relationship, and chlorophyll meter readings in soybean genotypes as affected by iron-ethylenediamine di-o-hydroxyphenylacetic acid in a calcareous soil. *J. Plant. Nutr.* **2007**, *31*, 81–89. [CrossRef]

50. Groenenberg, J.E.; Chardon, W.J.; Koopmans, G.F. Reducing phosphorus loading of surface water using iron-coated sand. *J. Environ. Qual.* **2013**, *42*, 250–259. [CrossRef] [PubMed]

51. Mei, X.Q.; Yang, Y.; Tam, N.F.Y.; Wang, Y.M.; Li, L. Roles of root porosity, radial oxygen loss, Fe plaque formation on nutrient removal and tolerance of wetland plants to domestic wastewater. *Water Res.* **2014**, *50*, 147–159. [CrossRef] [PubMed]

52. Geurts, J.J.M.; Smolders, A.J.P.; Verhoeven, J.T.A.; Roelofs, J.G.M.; Lamers, L.P.M. Sediment Fe:PO4 ratio as a diagnostic and prognostic tool for the restoration of macrophyte biodiversity in fen waters. *Freshw. Biol.* **2008**, *53*, 2101–2116. [CrossRef]

53. Bakker, E.S.; Van Donk, E.; Immers, A.K. Lake restoration by in-lake iron addition: A synopsis of iron impact on aquatic organisms and shallow lake ecosystems. *Aquat. Ecol.* **2016**, *50*, 121–135. [CrossRef]

54. Hossain, M.B.; Jahiruddin, M.; Loeppert, R.H.; Panaullah, G.M.; Islam, M.R.; Duxbury, J.M. The effects of iron plaque and phosphorus on yield and arsenic accumulation in rice. *Plant Soil* **2008**, *317*, 167–176. [CrossRef]

Article

Nitrate Attenuation in Degraded Peat Soil-Based Constructed Wetlands

Christian Kleimeier [1], Haojie Liu [1,*], Fereidoun Rezanezhad [2,3] and Bernd Lennartz [1,3]

[1] Faculty of Agricultural and Environmental Sciences, University of Rostock, Justus-von-Liebig-Weg 6, 18059 Rostock, Germany; kleimeier@hotmail.com (C.K.); bernd.lennartz@uni-rostock.de (B.L.)

[2] Ecohydrology Research Group, Water Institute and Department of Earth and Environmental Sciences, University of Waterloo, 200 University Avenue West, Waterloo, ON N2L 3G1, Canada; frezanez@uwaterloo.ca

[3] Baltic TRANSCOAST Research Training Group, Interdisciplinary Faculty, University of Rostock, 18059 Rostock, Germany

* Correspondence: haojie.liu@uni-rostock.de; Tel.: +49-0381-498-3193

Received: 20 February 2018; Accepted: 20 March 2018; Published: 22 March 2018

Abstract: Constructed wetlands (CWs) provide favorable conditions for removing nitrate from polluted agricultural runoff via heterotrophic denitrification. Although the general operability of CWs has been shown in previous studies, the suitability of peat soils as a bed medium for a vertical flow through a system for nitrate attenuation has not been proven to date. In this study, a mesocosm experiment was conducted under continuous flow with conditions aiming to quantify nitrate (NO_3^-) removal efficiency in degraded peat soils. Input solution of NO_3^- was supplied at three different concentrations (65, 100, and 150 mg/L). Pore water samples were collected at different depths and analyzed for NO_3^-, pH, and dissolved N_2O concentrations. The redox potential (Eh) was registered at different depths. The results showed that the median NO_3-N removal rate was 1.20 g/($m^2 \cdot$day) and the median removal efficiency was calculated as 63.5%. The nitrate removal efficiency was affected by the NO_3^- supply load, flow rate, and environmental boundary conditions. A higher NO_3^- removal efficiency was observed at an input NO_3^- concentration of 100 mg/L, a lower flow rate, and higher temperature. The results of pore water pH and NO_3^- and N_2O levels from the bottom of the mesocosm suggest that N_2 is the dominant denitrification product. Thus, degraded peat soils showed the potential to serve as a substrate for the clean-up of nitrate-laden agricultural runoff.

Keywords: nitrate attenuation; degraded peat; bed medium; constructed wetlands; mesocosm experiment

1. Introduction

Nitrogen (N) is an important nutrient in terrestrial and aquatic ecosystems. Around 120 million tons of nitrogen gases (N_2) per year are converted worldwide from the atmosphere into reactive nitrogen forms such as ammonium and nitrate by mineral fertilizer production and N-fixation by leguminous crops [1]. Most of the nitrogen reaching the terrestrial environment (directly or indirectly) is dissolved in surface runoff and infiltrating water. Regardless of the environmental compartment into which reactive N is released, much of the transported load ends up as NO_3-N in the aquatic environment [2]. Especially in lowland catchments with intensive agriculture, NO_3-N loads from agricultural fields may easily exceed 2000 kg/($km^2 \cdot$year), impacting drinking water quality [3]. In the German lowlands, the average NO_3-N losses from tile-drained field sites were found to vary between 340 and 2180 kg/($km^2 \cdot$year) [4].

Constructed wetlands (CWs) are often established as engineered systems to remove nitrate (i.e., denitrification) in wastewater or agricultural runoff by microbial denitrification and plant uptake [5].

Denitrification is an anaerobic respiration process in which nitrogenous oxides, principally NO_3^- and NO_2^-, are used as terminal electron acceptors and are hence reduced into the gaseous products nitric oxide (NO), nitrous oxide (N_2O), and dinitrogen (N_2) [6]. CWs generally consist of designed basins containing water, substrates, and plants. The type of substrates are important in CWs, because they provide storage capacity for contaminates via adsorption processes; in addition, many chemical and biological transformation processes occur at the surfaces of the various substrates. Wood products such as chips are commonly employed as denitrification beds in constructed wetlands due to the low cost and high C:N ratio (30:1 to 300:1, depending on the wood materials) [7–9]. Several other substrates, such as maize cobs and wheat straw, have also been proved suitable as denitrification beds [10]. Peat soils have a relatively low permeability and high organic carbon content, which will enhance the residence time of solutes or wastewater [11,12]. Thus, peat soils could potentially serve as denitrification beds for CWs. However, there are few studies about the performance of peat soils as denitrification bed materials [13–16].

Peatlands only cover 3% of the total global land area but store about one-third of the world's soil carbon. Drainage of peatland results in aeration and degradation of the peat, leading to the emission of greenhouse gases (e.g., CO_2) and losses of organic carbon [17]. For instance, greenhouse gas emissions of a peatland in Mecklenburg–Western Pomerania (northeast Germany) was estimated to be about 6.2 million tonnes of CO_2 equivalent per year [18]. Rewetting of peatlands is an effective practice to reduce greenhouse gas emissions, but may not affect nitrogen dynamics [19]. The uppermost highly degraded peat layers are recommended to be removed before any peatland restoration, because they will further release large amounts of organic compounds (e.g., dissolved organic carbon) into the downstream water bodies [20]. It has been reported that peat soils have a large amount of dead-end or closed pore volumes ranging from 40 to 80 vol % of total pore spaces [21–23]. Large numbers of immobile pores can be found in oxygen-depleted water regions, providing a favorable region for denitrification [15]. Highly degraded peat samples have been shown to have a high fraction of small (or immobile) pore regions [24], which can increase the denitrification activities. They can thus can be used as a suitable substrate for nitrate attenuation in CWs. This could be especially interesting in landscapes with scattered spots of fen peat. These local small-scale drained peatlands and wetlands have often been unsuccessfully converted into crops or grasslands, and may serve in the future as CWs for the cleanup of agricultural runoff [15,25].

In this study, we use a mesocosm experiment setup under controlled flux conditions to investigate the NO_3^- turnover processes in a lowland fen with highly degraded peat soils (soil organic matter content of 55% by weight) and examine the suitability of degraded peat for denitrification beds. The objectives of the study were: (1) to quantify the nitrate removal rate and efficiency under continuous flow conditions over a long-term period; (2) to explore the factors affecting the nitrate removal efficiency in peat soils; and (3) to evaluate the possibility of using degraded peat as filter medium in a constructed wetland to reduce nitrates from agricultural runoff.

2. Materials and Methods

2.1. Field Site and Sampling

The peat for the mesocosm experiment was collected from a small fen at Dummerstorf, 15 km southeast of the city of Rostock, Germany. It is a typical drained peatland in northern Germany with a highly degraded peat top soil. The soil carbon content of the top soil is 33.0% by weight [15,26]. The uppermost highly degraded peat was removed (0–20 cm depths) and transported to a greenhouse at the University of Rostock for the mesocosm experiment. The peat sample was homogenized manually (not sieved to maintain stable aggregates) using a clean rod, and any field-grown plant roots were removed. The peat material originated from a field, which is subjected to regular plowing; we thus assume that the mesocosm setup with homogenized peat material still reflects field conditions [15].

Homogenization was considered necessary to prevent the establishment of preferential flow paths and ensure the validity of the study.

2.2. Mesocosm Experimental Setup

To simulate a vertical flow through constructed wetland with open surface water, a container (polyethylene box) with dimensions of 100 cm length $\times$ 50 cm height $\times$ 30 cm width was used as experimental vessel (Figure 1). Drainage was facilitated through a 4-cm-thick layer of gravel at the bottom of the container, covered with a geo-textile membrane. Peat soil with field moisture content was filled into the container and the redox probes were installed vertically at different depths during the filling procedure. The depth of the peat in the mesocosm was 40 cm, with a total volume of 0.12 m^3. Bulk density was adjusted according to field conditions. The space above the peat was used to maintain a surface water table of 3 to 4 cm above the soil. The mesocosm container with peat soil was saturated from the bottom to minimize and remove the entrapped air from the peat, and during the experiments, the flow direction followed the gravimetric potential from top to bottom. The design of the flow through vessel corresponded to a constructed vertical flow wetland with an open water surface, as can be established in tile-drained landscapes where small scale fens often develop in local depressions.

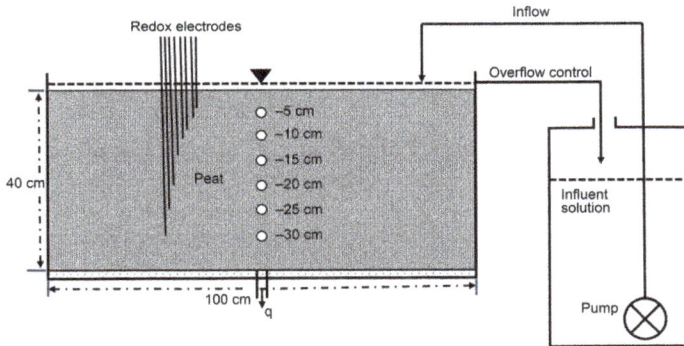

Figure 1. Schematic diagram of the mesocosm experimental set-up and illustration of water and gas sampling locations (−5, −10, −15, −20, −25, −30 cm, and effluent).

The input solution (pH = 7) contained 244 mg/L potassium nitrate (KNO$_3$) with three different NO$_3^-$ concentrations adjusted to 65, 100, and 150 mg/L. An NO$_3^-$ concentration of 65 mg/L (N$_{65}$) was chosen to represent the average annual NO$_3^-$ concentration in the ditch near the peat excavation site [27]. NO$_3^-$ concentrations of 100 mg/L (N$_{100}$) and 150 mg/L (N$_{150}$) represented two and three times the threshold given in the standard drinking water guidelines (50 mg/L [28]), respectively. The turnover experiment was conducted from April to August 2011. N$_{65}$ was firstly applied, application continued for 50 days, and then the N$_{100}$ was continuously applied for 28 days. After that, the concentration was shifted to N$_{150}$ and maintained for 42 days. Water sampling along the flow path was performed after the turnover experiment.

The flow rate (Q) for the entire experiment was maintained at about 24 liters per day (with a flux of about 8 cm/day; q = Q/A, A is the cross-sectional area of the bed container) and was adjusted by varying the hydraulic head. A self-constructed tipping counter, converting an effluent volume of 0.07 liter into an electric pulse, registered the flow rate. The redox potential Eh (mV) was measured at −5, −10, −15, −20, −25, and −30 cm depths below the peat surface by using custom-made 6-mm-diameter redox platinum electrodes (Sentec Ltd. Braintree, Essex, UK) connected to a Delta T DL2e data logger (Delta T Devices Ltd., Cambridge, UK). The measured values were corrected to the standard hydrogen potential E0 (mV) by adding the temperature-corrected voltage of the reference electrode according to the manufacturer's specifications. The temperature was additionally registered in the container with a Pt100 resistor-type temperature sensor (Ziehl industrie-elektronik GmbH, Schwaebisch Hall, Germany). More details about a similar mesocosm experimental setup can be found in Kleimeier et al. [15].

2.3. Pore Water Sampling

Six micro-piezometers (8-mm PVC pipes) were installed along the flow-path of the mesocosm for the pore water sampling at different depths. Piezometers were mounted horizontally into the peat-body in 5-cm intervals along the central flow path with Teflon gasket tape. The effluent sampling was performed automatically in 12-h intervals, using an ISCO 6712 automated portable sampler (Teledyne Isco Inc., Lincoln, NE, USA). The sampling ports were water- and gas-tight, not interfering with the adjusted flow rate.

The water and gas samples along the flow path were sampled with a 50-mL Braun Perfusor syringe (Carl Roth GmbH, Karlsruhe, Germany), 0.45-µm syringe-filters (Carl Roth GmbH, Karlsruhe, Germany), and a drip-infusion bag containing nitrogen gas. A syringe was connected to the piezometer (via a flexible extension tube, Carl Roth GmbH, Karlsruhe, Germany) to extract 50 mL of gas containing pore water. The sample (25 mL) was transferred into a second syringe, passing through the 0.45-µm filter. Immediately thereafter, 25 mL of N_2-gas were added to the syringe to create an anaerobic headspace via a 3-way luer-lock valve. The water samples were degassed by a 90-s ultrasonic treatment immediately after the sampling. The water sample was then transferred into 30-mL PE flasks, gassed with N_2 (to maintain the anaerobic state), and stored at 4 °C until analysis. The gas-phase was transferred into two 10-mL Exetainers (Labco, Ltd., Lampeter, Wales, UK), and analyzed for N_2O using a gas chromatograph (Shimadzu Europa GmbH, Kyoto, Japan). The content of dissolved gas was calculated using Henry´s law. In addition, for the water samples, nitrate (NO_3^-) concentrations were analyzed with an iron chromatograph (METROHM IC 700, Metrohm GmbH & Co. KG Filderstadt, Filderstadt, Germany).

2.4. Parameters and Statistical Analyses

The input NO_3-N_{load} and nitrate removal rate (NO_3-N_R) are given in consumed mass per unit of time and area, $g/(m^2 \cdot day)$.

$$NO_3 - N_{load} = C_{in} \times Q/(1000 \times A \times 4.43) \tag{1}$$

$$NO_3 - N_R = \Delta C \times Q/(1000 \times A \times 4.43) \tag{2}$$

where C_{in} is the inflow solution concentration (mg/L); Q is the flow rate (L/day); A is the cross-sectional area of the mesocosm system (m^2); and ΔC is the difference between the inflow and effluent (C_{out}) concentrations. The nitrate removal efficiency (N_{effi}) was calculated as follows:

$$N_{effi} = (NO_3 - N_R/NO_3 - N_{load}) \times 100\% \tag{3}$$

In order to derive a statistical model that represents the observed nitrate removal rate and efficiency as a function of environmental parameters (temperature, flow rate, and redox potential), the redox potential was grouped into three redox zones: zone 1, aerobic (>500 mV); zone 2, anoxic (150 mV < Eh < 500 mV); and zone 3, anaerobic (Eh < 150 mV) [15,29]. It was intended to calculate the thickness of the soil layer having anoxic or anaerobic conditions over the investigated time (cm/day, namely as the thickness and time duration of redox zone (TD_{zone}). Thus, TD_{zone1}, TD_{zone2}, and TD_{zone3} represent the thickness and time duration of redox zone 1, zone 2, and zone 3, respectively. Non-parametric Kruskal–Wallis [30] tests were performed to analyze whether nitrate removal rate, efficiency, and environmental parameters differed significantly among the three different nitrate input stages. To reveal the influence of nitrate load and environmental parameters on nitrate removal rate and efficiency, a stepwise linear regression analysis was conducted and all statistical analyses and modeling were performed using the default "stats" package of R [31].

3. Results

3.1. Nitrate Turnover

The amount of nitrate (NO_3-N) applied in the various stages of the experiment ranged from 0.30 to 3.91 g/(m^2·day) as determined on a daily basis analysis. The highest average input loading of 2.66 g/(m^2·day) NO_3-N was observed for the N_{150} experimental stage. The median NO_3-N removal rate over the whole experiment was measured as 1.20 g/(m^2·day) (Table 1). Removal efficiency of the entire mesocosm was 63.5%. The median value of difference between nitrate input and effluent concentrations (ΔC, C_{input}-C_{output}) was 64.0 mg/L. Although three nitrate concentrations (N_{65}, N_{100}, and N_{150}) were supplied into the mesocosm, there was no significant difference in nitrate removal rate between N_{65}, N_{100}, and N_{150}, whereas significances in nitrate load, NO_3-N removal efficiency, and ΔC were observed (Table 1). A higher ΔC was observed at N_{100}. The lowest nitrate removal efficiency occurred at N_{150} with an average value of 49.0%. With decreases in the applied nitrate concentration to N_{100} and N_{65}, the nitrate removal efficiency increased to 83.0% and 72.7%, respectively.

Table 1. Nitrate as nitrogen (NO_3-N) removal rate, removal efficiency and ΔC (difference between nitrate input and effluent concentrations), mean ± standard deviation, for different nitrate input concentrations: N_{65} (65 mg/L); N_{100} (100 mg/L); N_{150} (150 mg/L). Two groups that are identified by the same letter (a, b, or c) are not significantly different from each other.

Nitrate Input Concentrations	Input Loading	Removal Rate	Removal Efficiency	ΔC
	g/(m^2·day)	g/(m^2·day)	%	mg/L
N_{65}	1.72 b ± 0.90	1.19 a ± 0.57	72.7 a ± 16.1	47.2 b ± 10.5
N_{100}	1.46 b ± 0.50	1.15 a ± 0.30	83.0 a ± 18.2	83.0 a ± 18.2
N_{150}	2.66 a ± 0.69	1.22 a ± 0.34	49.0 b ± 10.9	73.3 a ± 16.2
Median	1.92	1.20	63.5	64.0

3.2. Environmental Factors

Temporal variation of the temperature and fluxes over the entire experimental period are illustrated in Figure 2. The mean volumetric flow was 25.8 L/day with a standard deviation of 12.9 L/day. The mean temperature was 26.2 °C and varied between daily average values of 15.5 °C and 38.8 °C over the 120 days of the experiment. A redox potential of less than 500 mV was observed at 5 cm soil depth and remained almost constant over the entire experiment period (Figure 2). Redox potentials at 5 to 15 cm depth ranged from 150 mV to 500 mV. Below a 30-cm depth, the redox potential remained under 150 mV over the observation period.

The temperature and flux varied between the three experimental stages of N_{65}, N_{100}, and N_{150} inputs (Figure 3). The flow rates at N_{65} were significantly higher than those at N_{100} and N_{150}, although we intended to maintain a constant flow rate. The average temperature for the N_{100} stage was 30.5 °C, which was higher than that at N_{65} (25.6 °C) and N_{150} (23.8 °C). No significant differences in the thickness and time duration of the redox zone Eh < 500 mV among N_{65}, N_{100}, and N_{150} were detected. For denitrification zones (150 mV < Eh < 500 mV and Eh < 150 mV), the lowest values of TD_{zone2} and TD_{zone3} were observed in N_{100} and N_{150} stages, respectively (Figure 3).

A stepwise linear regression analysis was conducted to delineate the individual specific effect of temperature and flux variations on nitrate turnover processes. The models with four parameters were developed to predict the nitrate removal rate and efficiency (Figure 4). A summary of these linear model results is given in Table 2. According to the r^2 statistic, the regression models revealed a 68% and 65% of the overall variance for estimation of nitrate removal rate and efficiency, respectively. In both models, most of the variability was attributable to the nitrate load (about 48%). The flux and temperature variations together accounted for approximately 20% of the total variation. Among them, the temperature contributes to 10% of the total variance.

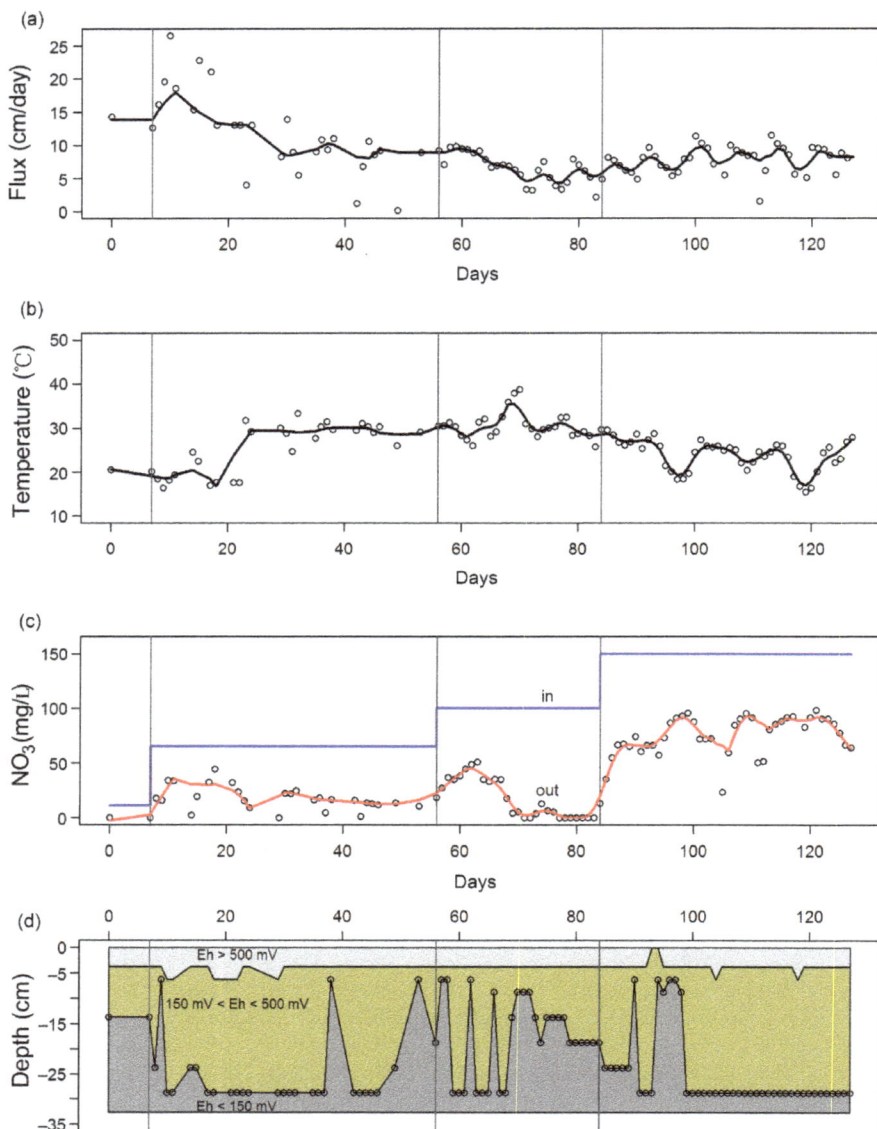

Figure 2. Overview of the flux (**a**), temperature (**b**), nitrate concentration (**c**), and redox potential zone (**d**) of the mesocosm experiment.

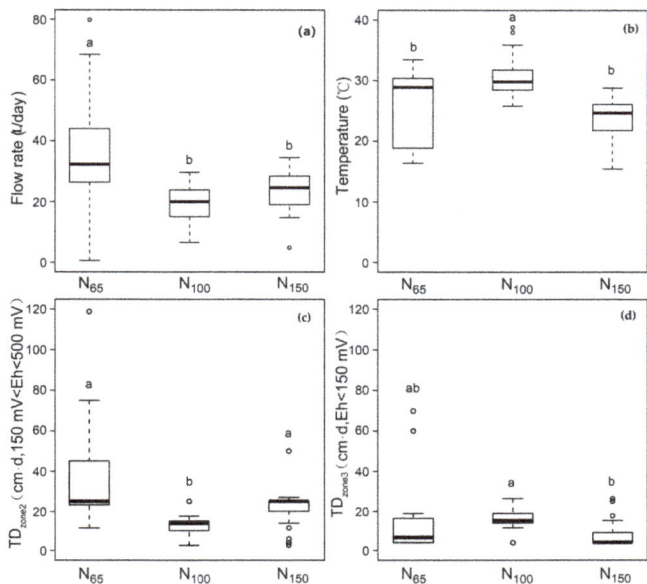

Figure 3. Boxplot of flow conditions, temperature, and thickness and time duration of redox zone (TD$_{zone}$) under different supplied nitrate concentrations of 65, 100, and 150 mg/L (N$_{65}$, N$_{100}$, N$_{150}$, respectively). Two groups that are identified by the same letter (a, b or c) are not significantly different from each other. (**a**) flow rate; (**b**) Temperature; (**c**) TD$_{zone2}$ (150 mV < Eh < 500 mV); (**d**) TD$_{zone3}$ (Eh < 150 mV).

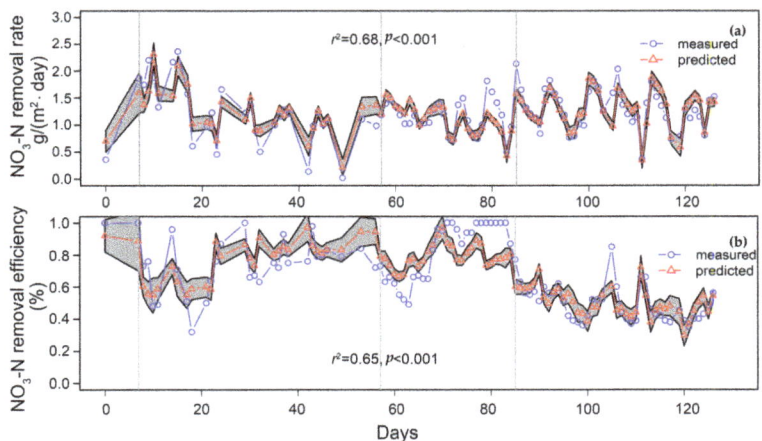

Figure 4. Measured (blue lines) and predicted (red line) values of nitrate removal rate and efficiency with multiple linear regression models. The grey zone is the 95% confidence of the interval of predicted values. (**a**) NO$_3$-N removal rate (**b**) NO$_3$-N removal efficiency.

Table 2. Multiple linear regression of nitrate as the nitrogen (NO_3-N) removal rate and efficiency with nitrate load, temperature, flow rate, and thickness and time duration of the redox zone (TD_{zone}). Level of significance ($p < 0.05$).

Index	Independent Variables	Coefficient	Explained Variance (%)
	Intercept	−0.800	-
	Input load	0.338	48.0
Removal rate, g/(m²·day)	Temperature	0.035	11.2
	Flow rate	0.014	6.8
	TD_{zone3} [a]	0.004	2.2
	Adjusted r^2	0.668	-
	Intercept	0.329	-
	Input Load	−0.152	46.8
Removal efficiency, %	Temperature	0.018	8.1
	Flow rate	0.006	8.7
	TD_{zone3}	0.002	2.3
	Adjusted r^2	0.645	-

[a] TD_{zone3}: Thickness and time duration (cm·day) of redox zone 3 (Eh < 150 mV).

4. Discussion

4.1. Nitrate Removal Rate and Efficiency

The nitrate removal rate of a biological denitrification process is strongly dependent on the multitude of possible C-sources for denitrification and the range of applied nitrogen loads. Nitrate removal is generally limited by the availability of C-sources [7]; adding an extra carbon source to denitrification bacteria in CWs could enhance the denitrification rate. In this study, a peat soil served as a C-source for the denitrification processes (C:N ratio of 10:1) [15]. The observed nitrate removal rates (1.21 g/(m²·day), NO_3-N) are significantly higher than those reported for CWs with mineral soils and natural wetlands (NO_3-N from 0.06 to 0.92 g/(m²·day)) [32,33]. The values were within nitrate removal rates found for woodchips (ranging from 0.7 to 5.0 g/(m²·day) NO_3-N) [7,34]. The obtained nitrate removal efficiency is higher than that reported for coco-peat (54%) and organic soils (20%) [16,35]. The reason probably was that lower oxygen concentrations (lower redox potential values) were observed in this study, especially at the bottom of the container. The nitrate removal rate at N_{100} was only a little lower than the rates estimated for wood-based filters (87% to 97%) [36]. However, at a later stage of the N_{100} variant, nitrate was almost completely removed (>95%), suggesting denitrification is limited by the nitrate concentration [36,37]. Thus, peat soils may have a comparable ability to wood products for removing nitrate in CWs.

4.2. Nitrate Load, Hydraulic Load, and Nitrate Removal Rate

Under nitrate-limiting conditions, an increasing nitrate input load would accelerate the nitrate removal process [38]. In this study, increasing the mean nitrate load from 1.46 g NO_3-N m⁻² day⁻¹ (N_{100}) to 2.66 NO_3-N g/(m²·day) (N_{150}) did not significantly improve the nitrate removal rate. We therefore assume that environmental factors (e.g., soil temperature) restrained the denitrification processes at N_{150}. When a nitrate load of less than 1.72 g/(m²·day) NO_3-N was supplied to the system, the nitrate effluent concentration (<10 mg/L NO_3-N or 50 mg/L NO_3) was always lower than the standard values for drinking water quality. However, at a larger nitrate load (e.g., 2.7 g/(m²·day)), the effluent concentration would often fall into the worst water quality class (>20 mg/L NO^{3-}N). For the setting in this study, the nitrate load should be limited to below 1.7 g/(m²·day) if degraded peat soil is used as the bed material to ensure water quality standards. In the regression models, the coefficient of the input load was positive for the nitrate removal rate but negative for nitrate removal efficiency because the nitrate removal rate increased less than the input load. Thus, removal efficiency decreased with increasing nitrate input load.

It has been reported that a higher hydraulic load commonly leads to high nitrate removal in CWs [39] because a higher flow rate often means a higher nitrate load and thus more nitrate available for denitrifying bacteria [40]. In this study, a positive relation between the hydraulic load and nitrate removal rate could be confirmed from the daily data set (Pearson's correlation coefficient of 0.67, $p < 0.01$). Our results are in line with a previous study [39], which stated that the nitrate removal rate is higher under a high hydraulic load condition. However, nitrate input load ($C_{input} \times Q$) rather than flow rate (Q) has a high negative correlation with removal efficiency (Pearson's correlation coefficient of 0.69, $p < 0.01$). Therefore, the status of an environmental system such as a catchment shall be evaluated on the basis of the total nutrient mass that has been released into the system. Lower ΔC values were observed for N_{65} at which the hydraulic load was highest (Figure 5a). Comparable ΔC values were obtained for N_{100} and N_{150}. We assumed that under a comparable nitrate load, faster flow rates would shorten the nitrate retention time in the system, resulting in a less effective transformation process [41]. For N_{150}, the output nitrate concentration always exceeds the standards for nitrate in drinking water (NO_3^-, 50 mg/L). Therefore, nitrate removal is very efficient up to an input concentration of 100 mg/L. When summarizing all data, a medium negative correlation between flow rate and ΔC (Figure 5b) indicates that the retention time plays an important role in the nitrate attenuation in peat.

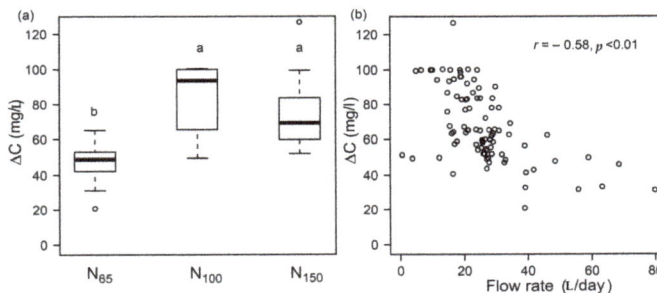

Figure 5. Boxplot of (**a**) ΔC (difference between nitrate input and effluent concentrations) under different supplied nitrate concentrations (N_{65}, N_{100}, N_{150}) and (**b**) scatter plot of flow rate (L/day) against ΔC.

4.3. Influence of the Temperature and Redox Potential on Nitrate Reduction

The nitrate removal efficiency is temperature- and oxygen-dependent [42–44] because higher temperature and oxygen availability increase the microbial activity in soils. In mineral soils, optimum temperatures for nitrification range from 15 to 35 °C [45] and for denitrification from 25 to 35 °C [46]. The highest nitrate removal efficiency was observed at the N_{100} period in which higher temperatures (about 30 °C) prevailed. Likewise, the regression model analysis clearly expressed the importance of temperature for the denitrification process in this study, where the temperature was more important than the extent and duration of oxygen-depleted zones (Table 2). A positive relationship was also found between nitrate removal efficiency and the TD_{zone} of denitrification zones, with redox potentials of less than 150 mV (Pearson's correlation coefficient of 0.31, $p < 0.01$). The thickness and time duration of the denitrification zone represent another factor affecting the nitrate removal efficiency, although they are less less important than temperature, flow rate, and input loading (Table 2).

4.4. Nitrate Turnover Along the Flow Path

Redox-sensitive solution sampling with a 5-cm resolution along the central flow path was conducted additionally to the regular automated effluent sampling. Measurements for nitrate and pH as well as the dissolved gas N_2O are shown in Figure 6. As expected, the redox potential decreases with increasing depth. Below a depth of 15 cm, the conditions are anaerobic. In general, nitrate concentrations were relatively high in the upper 25 cm. Lower nitrate concentrations were found closer to the bottom of the mesocosm in areas with lower redox potential indicating the bottom part of the container as the active denitrifying zone. The pH values ranged between 7.5 and 8, suggesting that N_2 is the dominant denitrification product [47]. The dissolved N_2O concentrations at all depths were low, confirming that the denitrifying bacteria convert nitrate to nitrogen gas (N_2) under anaerobic conditions [48].

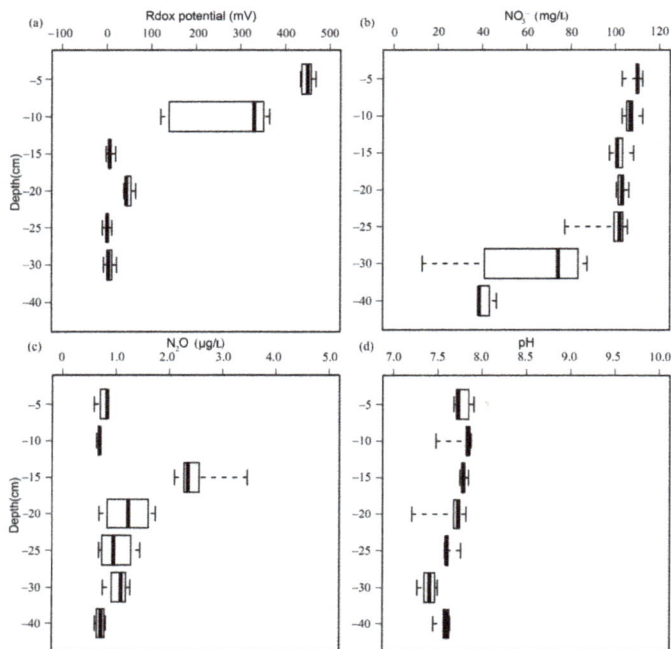

Figure 6. Redox potential (**a**), nitrate NO_3^- (**b**), and dissolved gasses N_2O (**c**) as well as the pH (**d**) at different depths in the mesocosm experiment.

5. Conclusions

In this study, we examined the suitability of degraded peat as a bed medium and C-source for nitrate removal in constructed wetlands. The nitrate removal rates and efficiency in degraded peat soils were comparable to those observed in woodchips reactors. The observed nitrate removal efficiency was strongly affected by the nitrate load and the environmental parameters, including flow rate, temperature, and redox potential. The highest removal efficiency (and ΔC) was obtained at N_{100}, with a lower flow rate and higher temperature. The lowest nitrate removal efficiency occurred at N_{150}, with an average value of 49.0%, where the nitrate removal efficiency increased to 83.0% and 72.7% with decreases in the applied nitrate concentration to N_{100} and N_{65}, respectively. Under a comparable nitrate input load, a continuous and low flow rate over a long-term period would extend the nitrate retention time, resulting in a more effective transformation process. The results showed that under a comparable nitrate load conditions, the nitrate retention time gets shorter under faster flow rates, resulting in a less effective nitrate transformation process. This paper clearly demonstrates the suitability of degraded peat as a bed medium for nitrate removal in constructed wetlands. For practical field-scale applications, the "flushing effect" of nitrate and dissolved organic carbon at early stages of the establishment of a CW should be taken into account to avoid unintended contamination of surface water bodies [15].

Acknowledgments: The European Social Fund (ESF) and the Ministry of Education, Science and Culture of Mecklenburg–Western Pomerania funded this work within the scope of the project WETSCAPES (ESF/14-BM-A55-0028/16). This study was conducted within the framework of the Research Training Group 'Baltic TRANSCOAST' funded by the DFG (Deutsche Forschungsgemeinschaft) under grant number GRK 2000 (www.baltic-transcoast@uni-rostock.de). This is Baltic TRANSCOAST publication no. GRK2000/0011. Christian Kleimeier is grateful to funding from the Department Maritime System of the University of Rostock.

Author Contributions: Christian Kleimeier and Haojie Liu contributed equally to this work. Christian Kleimeier designed the experiment; Haojie Liu and Christian Kleimeier conducted the experiment and analyzed the data; Haojie Liu wrote the original draft; Christian Kleimeier, Fereidoun Rezanezhad and Bernd Lennartz contributed in reviewing and editing the manuscript.

Conflicts of Interest: The authors declare no conflict of interest.

References

1. Rockström, J.; Steffen, W.; Noone, K.; Persson, Å.; Chapin, F.S.; Lambin, E.; Lenton, T.M.; Scheffer, M.; Folke, C.; Schellnhuber, H.J.; et al. A safe operating space for humanity. *Nature* **2009**, *461*, 472–475. [CrossRef] [PubMed]

2. Hey, D.L.; Kostel, J.A.; Crumpton, W.G.; Mitsch, W.J.; Scott, B. The roles and benefits of wetlands in managing reactive nitrogen. *J. Soil Water Conserv.* **2012**, *67*, 47A–53A. [CrossRef]

3. Syswerda, S.P.; Basso, B.; Hamilton, S.K.; Tausig, J.B.; Robertson, G.P. Long-term nitrate loss along an agricultural intensity gradient in the Upper Midwest USA. *Agric. Ecosyst. Environ.* **2012**, *149*, 10–19. [CrossRef]

4. Tiemeyer, B.; Lennartz, B.; Kahle, P. Analysing nitrate losses from an artificially drained lowland catchment (North-Eastern Germany) with a mixing model. *Agric. Ecosyst. Environ.* **2008**, *123*, 125–136. [CrossRef]

5. Lin, Y.F.; Jing, S.R.; Lee, D.Y.; Chang, Y.F.; Shih, K.C. Nitrate removal from groundwater using constructed wetlands under various hydraulic loading rates. *Bioresour. Technol.* **2008**, *99*, 7504–7513. [CrossRef] [PubMed]

6. Pell, M.; Wörman, A. Biological Wastewater Treatment Systems. In *Encyclopedia of Ecology*; Sven, E., Fath, B., Eds.; Academic Press: Oxford, UK, 2008; pp. 426–441.

7. Warneke, S.; Schipper, L.A.; Bruesewitz, D.A.; McDonald, I.; Cameron, S. Rates, controls and potential adverse effects of nitrate removal in a denitrification bed. *Ecol. Eng.* **2011**, *37*, 511–522. [CrossRef]

8. Vogan, J.L. The use of Emplaced Denitrifying Layers to Promote Nitrate Removal from Septic Effluent. Master's Thesis, University of Waterloo, Waterloo, ON, Canada, 1993.

9. Gibert, O.; Pomierny, S.; Rowe, I.; Kalin, R.M. Selection of organic substrates as potential reactive materials for use in a denitrification permeable reactive barrier (PRB). *Bioresour. Technol.* **2008**, *99*, 7587–7596. [CrossRef] [PubMed]

10. Cameron, S.G.; Schipper, L.A. Nitrate removal and hydraulic performance of organic carbon for use in denitrification beds. *Ecol. Eng.* **2010**, *36*, 1588–1595. [CrossRef]

11. Liu, H.; Forsmann, D.M.; Kjaergaard, C.; Saki, H.; Lennartz, B. Solute transport properties of fen peat differing in organic matter content. *J. Environ. Qual.* **2017**, *16*, 1106–1113. [CrossRef] [PubMed]

12. McCarter, C.P.R.; Price, J. The transport dynamics of chloride and sodium in a ladder fen during a continuous wastewater polishing experiment. *J. Hydrol.* **2017**, *549*, 558–570. [CrossRef]

13. Gunes, K. Restaurant Wastewater Treatment by Constructed Wetlands. *Clean-Soil Air Water* **2007**, *35*, 571–575. [CrossRef]

14. Amha, Y.; Bohne, H. Denitrification from the horticultural peats: Effects of pH, nitrogen, carbon, and moisture contents. *Biol. Fertil. Soils* **2011**, *47*, 293–302. [CrossRef]

15. Kleimeier, C.; Karsten, U.; Lennartz, B. Suitability of degraded peat for constructed wetlands—Hydraulic properties and nutrient flushing. *Geoderma* **2014**, *228–229*, 25–32. [CrossRef]

16. Jin, M.; Carlos, J.; McConnell, R.; Hall, G.; Champagne, P. Peat as Substrate for Small-Scale Constructed Wetlands Polishing Secondary Effluents from Municipal Wastewater Treatment Plant. *Water* **2017**, *9*, 928. [CrossRef]

17. Limpens, J.; Berendse, F.; Blodau, C.; Canadell, J.G.; Freeman, C.; Holden, J.; Roulet, N.; Rydin, H.; Schaepman-Strub, G. Peatlands and the carbon cycle: From local processes to global implications—A synthesis. *Biogeosciences* **2008**, *5*, 1475–1491. [CrossRef]

18. Mluv, M.V. Konzept zum Schutz und zur Nutzung der Moore. In *Fortschreibung des Konzeptes zur Bestandssicherung und zur Entwicklung der Moore*; Ministerium für Landwirtschaft, Umwelt und Verbraucherschutz Mecklenburg-Vorpommern: Schwerin, Germany, 2009; 109p.

19. Van Dijk, J.; Stroetenga, M.; Bos, L.; Bodegom, P.M.V.; Verhoef, H.A.; Aerts, R. Restoring natural seepage conditions on former agricultural grasslands does not lead to reduction of organic matter decomposition and soil nutrient dynamics. *Biogeochemistry* **2004**, *71*, 317–337. [CrossRef]

20. Zak, D.; Gelbrecht, J. The mobilisation of phosphorus, organic carbon and ammonium in the initial stage of fen rewetting (a case study from NE Germany). *Biogeochemistry* **2007**, *85*, 141–151. [CrossRef]

21. Rezanezhad, F.; Andersen, R.; Pouliot, R.; Price, J.S.; Rochefort, L.; Graf, M. How fen vegetation structure affects the transport of oil sands process-affected waters. *Wetlands* **2012**, *32*, 557–570. [CrossRef]

22. Rezanezhad, F.; Price, J.S.; Craig, J.R. The effects of dual porosity on transport and retardation in peat: A Laboratory Experiment. *Can. J. Soil Sci.* **2012**, *92*, 723–732. [CrossRef]

23. Rezanezhad, F.; Price, J.S.; Quinton, W.L.; Lennartz, B.; Milojevic, T.; Van Cappellen, P. Structure of peat soils and implications for water storage, flow and solute transport: A review update for geochemists. *Chem. Geol.* **2016**, *429*, 75–84. [CrossRef]

24. Kleimeier, C.; Rezanezhad, F.; Van Cappellen, P.; Lennartz, B. Influence of pore structure on solute transport in degraded and undegraded fen peat soils. *Mires Peat* **2017**, *19*, 1–9. [CrossRef]

25. Ketcheson, S.J.; Price, J.S.; Carey, S.K.; Petrone, R.M.; Mendoza, C.A.; Devito, K.J. Constructing fen peatlands in post-mining oil sands landscapes: Challenges and opportunities from a hydrological perspective. *Earth-Sci. Rev.* **2016**, *161*, 130–139. [CrossRef]

26. Rezanezhad, F.; Kleimeier, C.; Milojevic, T.; Liu, H.; Weber, T.K.D.; Van Cappellen, P.; Lennartz, B. The Role of Pore Structure on Nitrate Reduction in Peat Soil: A Physical Characterization of Pore Distribution and Solute Transport. *Wetlands* **2017**, *37*, 951–960. [CrossRef]

27. Tiemeyer, B.; Frings, J.; Kahle, P.; Köhne, S.; Lennartz, B. A comprehensive study of nutrient losses, soil properties and groundwater concentrations in a degraded peatland used as an intensive meadow—Implications for re-wetting. *J. Hydrol.* **2007**, *345*, 80–101. [CrossRef]

28. WHO. *Guidelines for Drinking-Water Quality*, 1st ed.; World Health Organisation: Geneva, Switzerland, 2004.

29. Sigg, L. Redox potential measurements in natural waters: Significance, concepts, and problems. In *Redox: Fundamental, Processes, and Applications*; Schuring, J., Schulz, H.D., Fischer, W.R., Bottcher, J., Duijnisveld, W.H.M., Eds.; Springer: Berlin, Germany, 2000.

30. McKight, P.E.; Najab, J. Kruskal-Wallis Test. In *Corsini Encyclopedia of Psychology*; John Wiley & Sons: Hoboken, NJ, USA, 2010.

31. R Core Team. *R: A Language and Environment for Statistical Computing*; R Foundation for Statistical Computing: Vienna, Austria, 2013; ISBN 3-900051-07-0.

32. Gale, P.M.; Dévai, I.; Reddy, K.R.; Graetz, D.A. Denitrification potential of soils from constructed and natural wetlands. *Ecol. Eng.* **1993**, *2*, 119–130. [CrossRef]

33. Blahnik, T.; Day, J. The effects of varied hydraulic and nutrient loading rates on water quality and hydrologic distributions in a natural forested treatment wetland. *Wetlands* **2000**, *20*, 48–61. [CrossRef]

34. Van Driel, P.W.; Robertson, W.D.; Merkley, L.C. Upflow reactors for riparian zone denitrification. *J. Environ. Qual.* **2006**, *35*, 412–420. [CrossRef] [PubMed]

35. Saeed, T.; Afrin, R.; Al Muyeed, A.; Sun, G. Treatment of tannery wastewater in a pilot-scale hybrid constructed wetland system in Bangladesh. *Chemosphere* **2012**, *88*, 1065–1073. [CrossRef] [PubMed]

36. Robertson, W.D.; Ford, G.I.; Lombardo, P.S. Wood-based filter for nitrate removal in septic systems. *Trans. ASAE* **2005**, *48*, 121–128. [CrossRef]

37. Schipper, L.A.; Robertson, W.D.; Gold, A.J.; Jaynes, D.B.; Cameron, S.C. Denitrifying bioreactors—An approach for reducing nitrate loads to receiving waters. *Ecol. Eng.* **2010**, *36*, 1532–1543. [CrossRef]

38. Almeida, A.; Carvalho, F.; Imaginário, M.J.; Castanheira, I.; Prazeres, A.R.; Ribeiro, C. Nitrate removal in vertical flow constructed wetland planted with *Vetiveria zizanioides*: Effect of hydraulic load. *Ecol. Eng.* **2017**, *99*, 535–542. [CrossRef]

39. Kadlec, R.H.; Knight, R.L. *Treatment Wetlands*; Lewis Publishers: Boca Raton, FL, USA, 1996.

40. Bastviken, S.K.; Weisner, S.E.B.; Thiere, G.; Svensson, J.M.; Ehde, P.M.; Tonderski, K.S. Effects of vegetation and hydraulic load on seasonal nitrate removal in treatment wetlands. *Ecol. Eng.* **2009**, *35*, 946–952. [CrossRef]

41. Gu, C.; Hornberger, G.M.; Mills, A.L.; Herman, J.S.; Flewelling, S.A. Nitrate reduction in streambed sediments: Effects of flow and biogeochemical kinetics. *Water Resour. Res.* **2007**, *43*, W12413. [CrossRef]

42. Willems, H.P.L.; Rotelli, M.D.; Berry, D.F.; Smith, E.P.; Reneau, R.B.; Mostaghimi, S. Nitrate removal in riparian wetland soils: Effects of flow rate, temperature, nitrate concentration and soil depth. *Water Res.* **1997**, *31*, 841–849. [CrossRef]

43. Abdalla, M.; Jones, M.; Smith, P.; Williams, M. Nitrous oxide fluxes and denitrification sensitivity to temperature in Irish pasture soils. *Soil Use Manag.* **2009**, *25*, 376–388. [CrossRef]

44. Veraart, A.J.; de Klein, J.J.M.; Scheffer, M. Warming Can Boost Denitrification Disproportionately Due to Altered Oxygen Dynamics. *PLoS ONE* **2011**, *6*, e18508. [CrossRef] [PubMed]

45. Lu, Y.; Xu, H. Effects of Soil Temperature, Flooding, and Organic Matter Addition on N_2O Emissions from a Soil of Hongze Lake Wetland, China. *Sci. World J.* **2014**, *2014*, 272684. [CrossRef] [PubMed]

46. Rivett, M.O.; Buss, S.R.; Morgan, P.; Smith, J.W.N.; Bemment, C.D. Nitrate attenuation in groundwater: A review of biogeochemical controlling processes. *Water Res.* **2008**, *42*, 4215–4232. [CrossRef] [PubMed]

47. Liu, B.; Mørkved, P.T.; Frostegård, Å.; Bakken, L.R. Denitrification gene pools, transcription and kinetics of NO, N_2O and N_2 production as affected by soil pH. *FEMS Microbiol. Ecol.* **2010**, *72*, 407–417. [CrossRef] [PubMed]

48. De Vries, S.; Schröder, I. Comparison between the nitric oxide re-ductase family and its aerobic relatives, the cytochrome oxidases. *Biochem. Soc. Trans.* **2002**, *30*, 662–667. [CrossRef] [PubMed]

water

MDPI

Article

Corn Straw as a Solid Carbon Source for the Treatment of Agricultural Drainage Water in Horizontal Subsurface Flow Constructed Wetlands

Yuanyuan Li, Sen Wang *, Yue Li, Fanlong Kong *, Houye Xi and Yanan Liu

College of Environmental Sciences and Engineering, Qingdao University, Qingdao 266071, China;
yuanyuanli0419@gmail.com (Y.L.); liyue@qdu.edu.cn (Y.L.); xihouye@gmail.com (H.X.);
liuyana0419@gmail.com (Y.L.)
* Correspondence: wangsen@qdu.edu.cn (S.W.); kongfanlong@qdu.edu.cn (F.K.)

Received: 28 February 2018; Accepted: 18 April 2018; Published: 20 April 2018

Abstract: Agricultural drainage water with a low C/N ratio restricts the nitrogen and phosphorus removal efficiencies of constructed wetlands. Thus, there is a need to add external carbon sources to drive the nitrogen and phosphorus removal. In this study, the effects of the addition of corn straw pretreated with different methods (acid treatment, alkali treatment, and comminution) on treating agricultural drainage water with a low C/N ratio were investigated in constructed wetlands. The results showed that soaking the corn straw in an alkaline solution was the most suitable pretreatment method according to the release rule of chemical oxygen demand (COD) and the dissolution of total nitrogen (TN) and total phosphorus (TP). The average removal efficiency of TN and TP in constructed wetlands increased respectively by 37.2% and 30.5% after adding corn straw, and by 17.1% and 11.7% after adding sodium acetate when the hydraulic retention time (HRT) was 3 days. As an external carbon source, straw was cheap, renewable, and available. In contrast, the sodium acetate demanded high costs in a long-term operation. Therefore, corn straw had a great advantage in treatment effect and cost, which improved the treatment efficiency of agricultural drainage water using a byproduct of agricultural production as a slow-release carbon source.

Keywords: corn straw; solid carbon sources; agricultural drainage water; low C/N ratio; constructed wetland

1. Introduction

More and more fertilizers have been widely used to increase the agricultural production and alleviate the food crisis caused by population growth. However, due to the low utilization rate, large amounts of unutilized fertilizers run into the surface water along with agricultural drainage water [1]. The agricultural drainage water often contains COD, organic nitrogen, NH_4^+-N, NO_3^--N, and some inorganic phosphates [1]. It has the characteristics of a low C/N ratio, high proportion of nitrate, and the fluctuation of water quality and quantity [2]. The discharged drainage water can easily result in eutrophication [3]. Therefore, an economical and practical treatment technique for agricultural drainage water is desired.

The low C/N ratio in wastewater is often treated with processes such as anaerobic ammonium oxidation and simultaneous nitrification and denitrification, which have the shortcomings of complex operation and management, high energy consumption, and high cost [4,5]. Constructed wetlands (CWs) are known as an ecological technology with a good purification effect, simple process equipment, low setup and maintenance costs, and high ornamental value [6]. CWs have been successfully applied to treat domestic, municipal, and industrial wastewater, as well as contaminated surface water by combining physical, chemical, and biological processes [7–9]. It is well known that biological nitrogen removal mainly relies on successful ammonia oxidation and nitrate denitrification to nitrogen gas. In the absence of a carbon source, the denitrification will be restricted in CWs [10]. Therefore, an additional carbon source is necessary for treating agricultural drainage water with a low C/N ratio. Studies have shown that traditional liquid carbon sources, including glucose, methanol, ethanol, starch, and sodium acetate, can be used to improve the efficiency of biological denitrification in CWs [11,12]. The liquid carbon source needs to be added continuously, which is high-cost and can easily cause secondary pollution [13]. Compared with liquid carbon sources, solid carbon sources can not only work for a long time, but also create a stable living environment for the denitrifying bacteria [14]. Recent research has indicated that low-cost and renewable natural organic substances, such as corn straw, rice husk, litter, woodchips, sawdust, cotton, maize cobs, seaweed, and bark, can be used as external solid carbon sources to drive denitrification [15]. Shao et al. [16] demonstrated that rice husks can be an economical and effective carbon source and biofilm carrier in the biological denitrification of wastewater in up-flow laboratory reactors. Chen et al. [17] reported that *Typha latifolia* litter addition could greatly improve nitrate removal in subsurface-batch CWs through the continuous input of organic carbon. Yang et al. [18] found that *retinervus luffae fructus*, corncob, and rice straw were the favorable solid carbon sources and biofilm carriers due to their better carbon release capacity, denitrification potential, and relatively large surface area. Compared to the control membrane bioreactors, TN removal was enhanced by 25.5%, 19.5%, and 38.9%, respectively. Xu et al. [19] and Li et al. [20] also successfully drove denitrification using corncobs. In addition, corn straw, *Arundo donax*, and *Pontederia cordata* were also proven to be effective carbon sources [11,21]. However, few reports have been reported on the treatment of agricultural drainage water using corn straw as carbon source in CWs, as well as the influence of pretreatment methods on the carbon source release of corn straw.

In this study, common corn straw was selected as the solid carbon source. The objectives of this study were: (1) to analyze the effects of pretreatment methods on the amount and rate of carbon released by corn straw; (2) to study the contribution of the solid carbon source to the removal of COD, N, and P in the CW system; and (3) to assess the economy of corn straw used as a slow-release solid carbon source in treating agricultural drainage water with a low C/N ratio.

2. Materials and Methods

2.1. Pretreatment of Corn Straw

Natural corn straw used in all CWs of this study was collected from a local village of Chiping county in Shandong province. The corn straw was cleaned with tap water before drying in the air, cut into 2–3-cm pieces, and then dried at 40 °C to a constant mass. The treated corn straw was kept in a moisture-free container before use.

2.2. Carbon Dissolution Test

Corn straw was treated with four different processes to explore the effects of pretreatment methods (Figure 1). After pretreatment, 2.0 g corn straw was put into 100 mL distilled water and soaked under the water. The collected water samples (once per hour at first and then once a day) were immediately filtered through a 0.45-μm cellulose acetate membrane and analyzed for the dissolution of COD, TN, and TP.

2.3. Characteristics of the Wastewater

Based on the characteristics of agricultural drainage water in different areas [22–24], synthetic wastewater was prepared. The composition of the synthetic wastewater was as follows (mg/L): NH_4Cl 5.730, KH_2PO_4 6.590, $C_6H_{12}O_6$ 26.416, and KNO_3 75.825. This solution represented 28.0 mg/L COD, 1.5 mg/L NH_4^+-N, 10.5 mg/L NO_3^--N, 12.0 mg/L TN, and 1.5 mg/L TP. The pH of the synthetic wastewater was 7.2–7.5.

Figure 1. (**a**) Non-treated corn straw; (**b**) corn straw treated with acid (solid-liquid ratio 1:20, soaked in 20% HCl for 24 h, washed, and then dried to constant weight); (**c**) corn straw treated with alkali (solid-liquid ratio 1:30, soaked in 1% NaOH for 24 h, washed, and then dried to constant weight); and (**d**) comminuted corn straw.

2.4. Wetland Setup and Operation

The five sets of systems established in the greenhouse at Qingdao University were as follows: no added carbon source (W1: HRT = 2 days), added corn straw (W2: HRT = 2 days), added sodium acetate (W3: HRT = 2 days), added corn straw (W4: HRT = 3 days), and added corn straw (W5: HRT = 1 day). The volume of each system was 25 L (length: 0.45 m, width: 0.30 m, height: 0.30 m). All CWs reactors were filled with washed gravel and sand. The top of the CWs were open and the bottom had one opening valve for emptying. In five CWs, multi-dimensional gradation of the gravel was adopted: larger sized gravels (5–8 cm) were placed at the bottom to avoid clogging, on top of which were smaller sized gravels (1–5 cm). Finally, a 5-cm layer of washed sand (particle size < 2 mm) was placed at the top, in which the *Typha latifolia* was planted. Continuous experiments were carried out in an air-conditioned greenhouse at 23 ± 1 °C. After planting, the systems were operated for four weeks with synthetic wastewater until the plant shoots and microorganisms were well established. According to previous research, we chose a proper C/N ratio to achieve a complete denitrification (C/N = 8) [6,25,26]. About 1.87 g sodium acetate was continuously added to W3 daily. Based on the results of the carbon source release, 68.08, 43.57, and 136.16 g corn straw was added to W2, W4, and W5, respectively.

2.5. Sampling and Analysis

The water samples were collected from the reactor and filtered through a 0.45-μm cellulose acetate membrane before analysis. The pH and temperature were monitored by a pH/mV Meter (PHS-3CW, Bante Instrument Co. Ltd., Shanghai, China) and Pen type thermometer (TP101, Shanghai Automation Instrument Factory, Shanghai, China), while conventional pollutants in wastewater (COD, NH_4^+-N, NO_3^--N, TN, and TP) were determined according to Chinese standard methods [27]. COD was measured using the potassium dichromate method. NH_4^+-N, NO_3^--N, TN, and TP were determined by a UV–vis spectrophotometer (TU-1810, Persee instrument Co. Ltd., Beijing, China) [28].

2.6. Statistical Analysis

All tests in this study were performed in triplicate and the results are expressed as mean ± standard deviation. Experimental data was processed using Microsoft Excel 2013 and the charts were drawn using Microsoft Excel 2013 and Origin 8.0 software.

3. Results and Discussion

3.1. Effect of Pretreatment Methods on the Carbon Source Release of Corn Straw

As shown in Figure 2, the carbon, nitrogen, and phosphorus release rates of the four samples were relatively fast at the early stage, after which they tended to be stable. The COD contents released by four kinds of corn straw were respectively 55.0, 67.5, 94.8, and 242.5 mg/g at the early stage (Figure 2a), due to the release of small molecule organic matter attaching to the surface of the corn straw. In particular, the comminuted corn straw with a larger surface area released a much higher COD content than other treatments. After about 4 days, the small molecule organic matter in the surface was released completely, and most of the COD in water samples were from cellulosic materials, which is main component of corn straw [29]. Alkali treatment was mainly applied to cellulose, which was hydrolyzed to glucose and other monosaccharides, while acid treatment acted mainly on nucleoside bonds that caused fragmentation [30]. Therefore, the released COD contents levels were in the order of c > b > a > d, corresponding to 6.24, 4.17, 2.49, and 1.01 mg/g, respectively. At the early stage, the TN contents released by four kinds of corn straw were in order of d > a > b > c (Figure 2b), and the TP contents were in the order of d > b = c > a (Figure 2c). This was mainly due to the fact that the nitrogen and phosphorus contents in corn straw were high, and the release of nitrogen and phosphorus sped up along with the rapid decomposition of organic matter. The release of nitrogen and phosphorus was similar to the results of previous researchers [1,29]. After 6 days, the TN contents from a, b, c, and d were 0.272, 0.291, 0.233, and 0.266 mg/g, and the TP contents were 0.005, 0.003, 0.003, and 0.005 mg/g, respectively, which were very close to each other. The above results suggested that different treatment methods had no significant effect on the nitrogen and phosphorus release from corn straw.

The C/N values of a, b, c, and d exhibited a sharp decline in the early stage, then rose and finally tended to be stable, at levels of 9.4, 11.8, 24.3, and 4.0, respectively (Figure 3). It was obvious that the corn straw treated by alkali could provide more carbon and little nitrogen as an external carbon source.

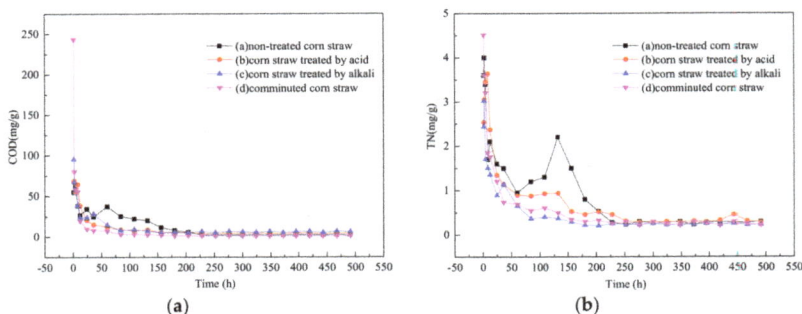

Figure 2. *Cont.*

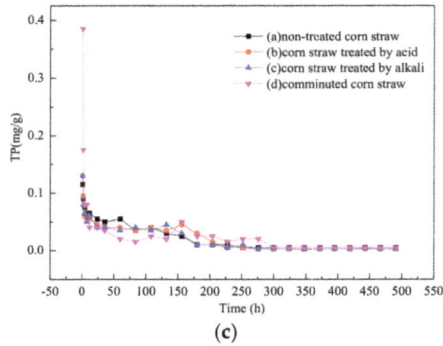

(c)

Figure 2. COD, TN, TP dissolution trends of corn straw pretreated with different methods; (a) COD; (b) TN; and (c) TP.

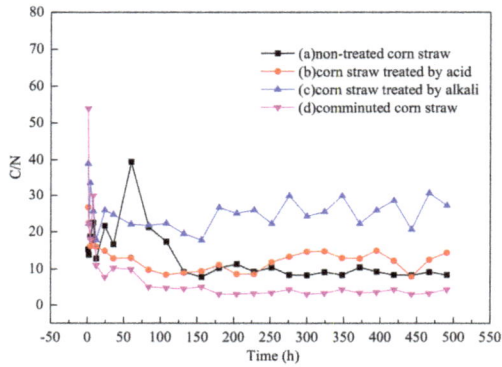

Figure 3. C/N variation trends of corn straw pretreated with different methods.

3.2. Effects of External Carbon Source and HRT on COD Removal

The five CWs reached a steady state after being cultivated with synthetic wastewater for four weeks (Table 1). After the CWs were stable, an external carbon source was added to W2, W3, W4, and W5. As shown in Figure 4, the influent COD concentrations of W1 and W2 were 26.40–29.30 mg/L and the average COD removal efficiency was 63.2% and 45.0%, respectively. The influent COD concentration of W3 was 95.60–98.60 mg/L, and the average COD removal efficiency was 74.5%. This indicated that the CWs had a low effluent COD concentration after adding corn straw and sodium acetate. The effect of HRT on the COD removal is shown in Figure 4. The influent COD concentrations of W2, W4, and W5 were in the range of 26.40–29.30 mg/L and the average COD removal efficiencies were 45.0%, 54.1%, and 40.7%, respectively, indicating that the HRT had an impact on COD removal in CWs, and that the optimum HRT was 3 days.

Table 1. The removal efficiency of COD, NH$_4^+$-N, NO$_3^-$-N, TN, and TP in five constructed wetlands before adding external carbons.

Reactor	COD			NH$_4^+$-N			NO$_3^-$-N			TN			TP		
	Influent Content [a] (mg/L)	Effluent Content [a] (mg/L)	Average Removal Efficiency (%)	Influent Content [a] (mg/L)	Effluent Content [a] (mg/L)	Average Removal Efficiency (%)	Influent Content [a] (mg/L)	Effluent Content [a] (mg/L)	Average Removal Efficiency (%)	Influent Content [a] (mg/L)	Effluent Content [a] (mg/L)	Average Removal Efficiency (%)	Influent Content [a] (mg/L)	Effluent Content [a] (mg/L)	Average Removal Efficiency (%)
W1	27.0 ± 1.0	10.0 ± 1.0	61.4	1.50 ± 0.10	0.60 ± 0.15	61.4	10.5 ± 0.6	4.50 ± 0.20	57.1	12.0 ± 0.5	5.80 ± 0.2	50.7	1.50 ± 0.15	0.80 ± 0.10	35.4
W2	28.0 ± 1.5	13.0 ± 1.7	54.4	1.50 ± 0.25	0.50 ± 0.10	65.6	11.0 ± 0.6	4.28 ± 0.13	60.0	12.0 ± 0.4	4.78 ± 0.2	53.2	1.50 ± 0.15	0.75 ± 0.10	37.5
W3	28.3 ± 1.3	12.1 ± 2.1	55.6	1.55 ± 0.10	0.60 ± 0.30	58.1	10.7 ± 1.0	4.07 ± 0.23	61.6	12.3 ± 1.0	4.67 ± 0.3	54.4	1.55 ± 0.10	0.75 ± 0.10	38.3
W4	27.6 ± 0.8	9.1 ± 2.0	67.5	1.60 ± 0.15	0.50 ± 0.05	65.3	11.0 ± 0.5	3.57 ± 0.50	68.5	12.6 ± 0.5	4.07 ± 0.4	64.2	1.45 ± 0.10	0.65 ± 0.10	56.7
W5	26.6 ± 1.8	13.4 ± 1.9	48.4	1.60 ± 0.10	0.80 ± 0.10	49.3	11.0 ± 0.5	5.40 ± 0.40	50.1	12.6 ± 0.5	6.20 ± 0.5	36.8	1.45 ± 0.10	0.95 ± 0.20	30.5

[a] Values are mean values ± standard deviations.

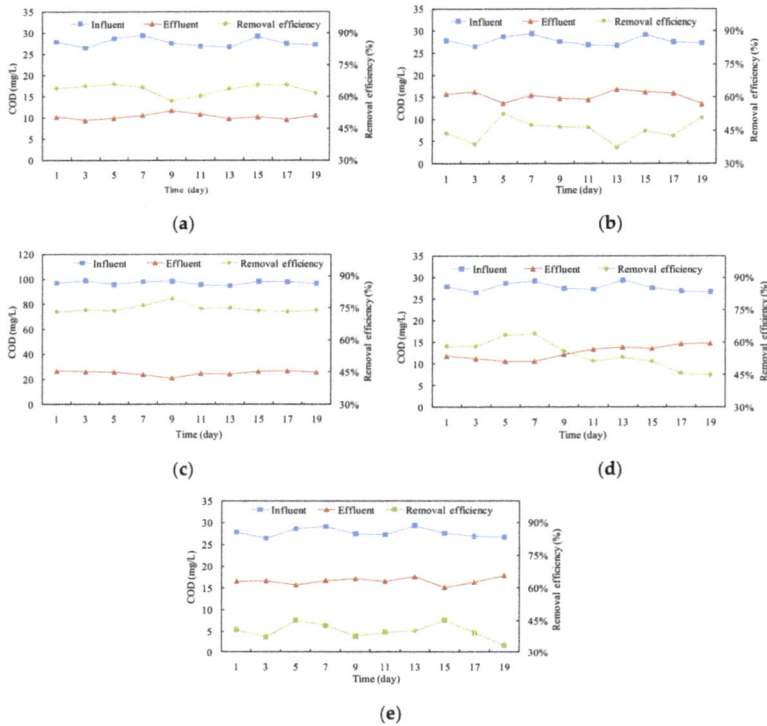

Figure 4. COD removal efficiency in five constructed wetlands (CWs); (**a**) W1, non-added, HRT = 2 days; (**b**) W2, added corn straw, HRT = 2 days; (**c**) W3, added sodium acetate, HRT = 2 days; (**d**) W4, added corn straw, HRT = 3 days; and (**e**) W5, added corn straw, HRT = 1 day.

3.3. Effects of External Carbon Source and HRT on Nitrogen Removal

The effect of the external carbon source on NH_4^+-N removal is shown in Figure 5a. The influent NH_4^+-N concentrations of W1, W2, and W3 were in the range of 1.45–1.55mg/L, and the average NH_4^+-N removal efficiencies were respectively 67.2%, 74.8%, and 75.2%, indicating that the external carbon source improved the NH_4^+-N removal and that corn straw was more efficient than sodium acetate. The first reason for this was that the denitrifying bacteria grew vigorously in enough carbon sources, which also required a certain amount of nitrogen during growth. Secondly, the CW was a mixed culture system, where nitrifying bacteria, autotrophic bacteria, and denitrifying bacteria coexisted. Therefore, nitrification, denitrification, and anaerobic ammonia oxidation might occur simultaneously [11]. In W2, the corn straw pretreated by alkali might release ammonia while providing a carbon source. Thus, the NH_4^+-N in W2 was slightly larger than that in W3.

The NO_3^--N concentrations of the influent in W1, W2, and W3 were in the range of 9.90–11.40 mg/L, and the average effluents were 4.89, 1.69, and 1.72 mg/L, respectively (Figure 5b). In W1, denitrifying bacteria could not obtain a sufficient carbon source, so nitrogen removal efficiency was poor. However, denitrifying bacteria could obtain enough carbon sources in W2 and W3 after the addition of an external carbon source. Therefore, an external carbon source can improve the denitrification efficiency of CWs [17,31]. Corn straw showed a better promotion than sodium acetate, because the corn straw could not only increase the available carbon source, but also provide places for microbial attachment, thus increasing the biomass [32,33]. Moreover, the influent TN concentrations in W1, W2, and W3 were in the range of 11.35–12.90 mg/L, and the average TN removal efficiencies were 50.8%, 71.9%, and 67.9%, respectively (Figure 5c). It was obvious that the nitrification and denitrification progress were promoted in W2 and W3 after adding the external carbon source. However, the removal efficiency in W2 was 4.0% higher than that in W3. This indicated that as a solid slow-release carbon source, corn straw was a better choice for treating low C/N wastewater in CWs than sodium acetate.

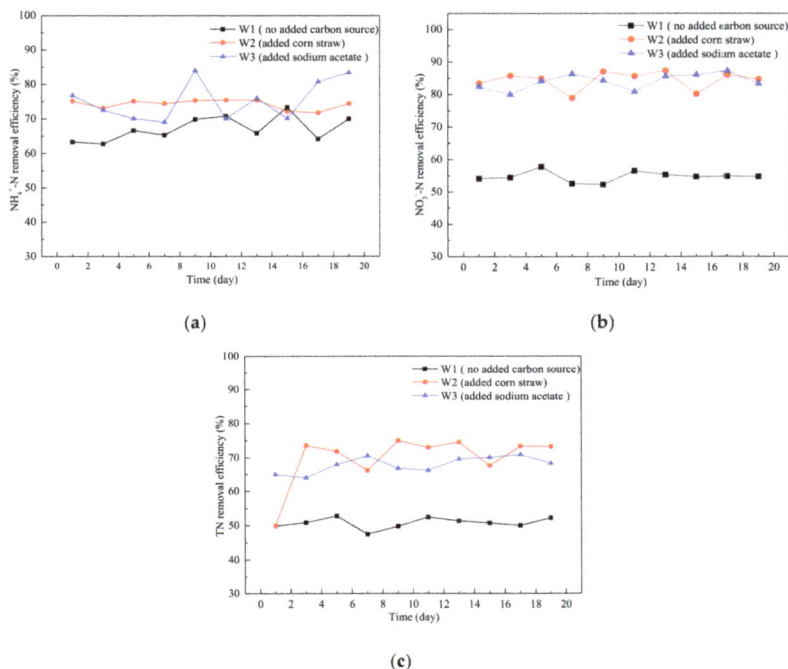

(a)

(b)

(c)

Figure 5. Effects of external carbon source on nitrogen removal; (**a**) NH_4^+-N; (**b**) NO_3^--N; and (**c**) TN.

HRT had an influence on nitrogen removal. The NH_4^+-N concentrations in W2, W4, and W5 were in the range of 1.45–1.55 mg/L; and the average removal efficiencies were 74.8%, 79.3%, and 65.6%, respectively (Figure 6a). The above result indicated that HRT had an influence on NH_4^+-N removal, and the NH_4^+-N removal efficiency in W4 (HRT = 3 days) was significantly higher than those in W2 and W5. The HRT of 3 days was optimum for microbial nitrification, which meant better NH_4^+-N removal. The influent NO_3^--N concentrations of W2, W4, and W5 were in the range of 9.90–11.40 mg/L, and the effluent NO_3^--N concentrations were 1.69, 0.72, and 3.65 mg/L, respectively (Figure 6b). This suggested that the HRT had a great influence on denitrification [34]. In addition, the TN removal efficiencies in W2, W4, and W5 varied with the HRT, which was 71.9%, 87.9%, and 52.3%, respectively (Figure 6c). The HRT of 3 days was also optimum for the removal of TN in CWs [35].

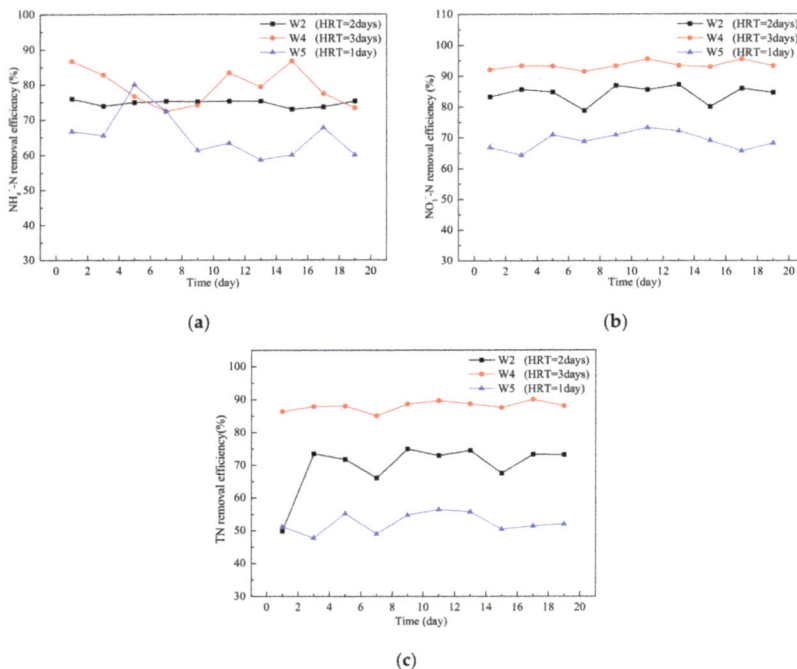

Figure 6. Effect of HRT on NH_4^+-N, NO_3^--N, TN removal; (**a**) NH_4^+-N; (**b**) NO_3^--N; and (**c**) TN.

3.4. Effects of External Carbon Source and HRT on Phosphorus Removal

The phosphorus removal in CWs includes biotic processes (e.g., the uptake and growth of plants and microorganisms) and abiotic processes (e.g., settling, sorption on substrates, co-precipitation with minerals, adsorption, and precipitation) [36,37]. Phosphorus is mostly removed through precipitation/ adsorption in the media. Plant uptake and biological assimilation are limited processes. Since the CWs were relatively new, the increase in phosphorus removal observed with the addition of the carbon source was attributed to the faster build-up of biofilm within the system and to the clean and unsaturated gravel media. In this study, the phosphorus removal through the uptake and growth of microorganisms was different in the same five CW systems. In the aerobic zone, phosphate-accumulating organisms (PAOs) could oxidize organics material to obtain energy, resulting in an increase of phosphate concentration in the water. Under aerobic conditions, excessive phosphorus accumulated by PAOs was converted into polyphosphate using oxygen as an electron acceptor and then stored inside the cells.

As shown in Figure 7a, the influent TP concentrations in W1, W2, and W3 were in the range of 1.40–1.55 mg/L, and the average TP removal efficiencies were 34.9%, 49.7%, and 46.6%, respectively. For agricultural drainage water with a low C/N ratio, the release of phosphorus was suppressed. Therefore, W2 and W3 were better than W1 in the treatment of phosphorus after the addition of a carbon source. The TP removal efficiencies in W2, W4, and W5 varied with the HRT, which were 49.7%, 65.4%, and 33.4%, respectively (Figure 7b). Phosphorus removal in the CWs consisted of physical and chemical reactions, plant uptake, and microbial assimilation. With the increasing HRT, the above progresses played out more fully. Thus, the HRT of 3 days was optimum for the removal of TP in CWs.

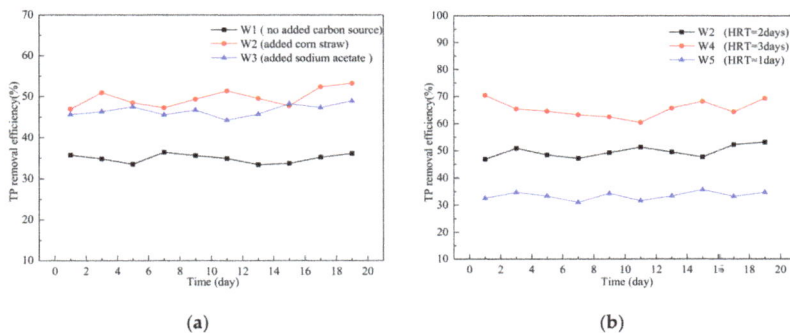

Figure 7. Effects of carbon source and HRT on TP removal: (**a**) carbon source; (**b**) HRT.

3.5. The Cost Analysis of Different External Carbon Sources

The average removal efficiencies of TN after adding corn straw and sodium acetate were 71.9% and 67.9%, respectively. It was obvious that corn straw and sodium acetate were good choices as external carbon sources to remove nitrogen from the wastewater. The cost of corn straw and sodium acetate as the external carbon source was calculated according to the wastewater treatment capacity of 1 t/day. One liter of agricultural drainage water required 0.14 g sodium acetate, which was about 2.20 RMB/kg. The cost of sodium acetate was about 43.20 RMB for 20 days. Conversely, 1 L agricultural drainage water needed 5.30 g corn straws, costing little money. Before use, the corn straw needed to be pretreated with NaOH, requiring about 3.30 RMB/kg. The cost of corn straw and NaOH was about 10.36 RMB for 20 days, which was lower than that of sodium acetate under the same conditions.

In the north of China, corn is a common crop. As a byproduct of corn, corn straw is widely abundant and its price is very low (about 300.00 RMB/t). Therefore, enough corn straw can be provided for full-scale CWs. Corn straw used in full-scale wetland systems need to be purchased, transported, pretreated, and distributed. The following cost calculation was conducted according to the wastewater treatment capacity of 100 t/day. The overall cost was about 1105.62 RMB, including the purchasing cost of 190.80 RMB, transportation cost of 200.00 RMB, pretreatment cost of 314.82 RMB, and labor cost of 400.00 RMB. Given that corn straw can efficiently work 30 days at least, the treating cost of a ton of waste water is about 0.37 RMB. Due to the low cost, sufficient quantity, and good effect, it is feasible to use corn straw as an external carbon source in CWs.

4. Conclusions

This study investigated the effects and economy of using corn straw as an external carbon source on treating agricultural drainage water with a low C/N ratio. Through different pretreatments (acid treatment, alkali treatment, and comminution), some organic matter, N and P, was released from the corn straw in the dissolution process. Compared with the other two pretreatment methods, the alkali treatment could provide more COD with lower N and P release by hydrolyzing the cellulose material of corn straw to glucose and other monosaccharides. The removal efficiencies of TN and TP were significantly promoted by adding straw and sodium acetate as the external carbon sources, compared with the control test. The effects of the HRT on the removal of N and P were also studied. It was revealed that the optimum HRT was 3 days. As an agricultural byproduct, the corn straw was cheap and easily obtained, which presented great advantages in treating agricultural drainage water with a low C/N ratio in CW. The long-term effects of corn straw as an external carbon source on treating agricultural drainage water in CWs will be studied in the future.

Acknowledgments: This work was supported by the National Natural Science Foundation of China (41771098).

Author Contributions: Yuanyuan Li and Sen Wang conceived and conducted the experiments under the supervision of Yue Li and Fanlong Kong. Yuanyuan Li, Houye Xi, and Yanan Liu analyzed the data, and Yuanyuan Li and Sen Wang wrote the paper. Yue Li, Fanlong Kong, and Sen Wang provided critical revision of the manuscript. All authors read and approved the submitted manuscript.

Conflicts of Interest: The authors declare no conflict of interest.

References

1. Zhao, J.; Zhao, Y.; Zhao, X.; Jiang, C. Agricultural runoff pollution control by a grassed swales coupled with wetland detention ponds system: A case study in Taihu Basin, China. *Environ. Sci. Pollut. Res.* **2016**, *23*, 9093–9104. [CrossRef] [PubMed]

2. Chang, Y. Nitrogen Removal Efficiency of the Carbon-Sulfur Coupling Surface Flow Constructed Wetland. Master's Thesis, Chang'an University, Xi'an, China, 2016.

3. Hua, Y.; Peng, L.; Zhang, S.; Heal, K.V.; Zhao, J.; Zhu, D. Effects of plants and temperature on nitrogen removal and microbiology in pilot-scale horizontal subsurface flow constructed wetlands treating domestic wastewater. *Ecol. Eng.* **2017**, *108*, 70–77. [CrossRef]

4. Chen, Y.; Peng, Y.; Wang, J. Biological phosphorus and nitrogen removal in low C/N ratio domestic sewage treatment by A^2/O-BAF combined system. *Acta Sci. Circumst.* **2010**, *30*, 1957–1963.

5. Lee, H.; Han, J.; Yun, Z. Biological nitrogen and phosphorus removal in UCT-type MBR process. *Water Sci. Technol.* **2009**, *59*, 2093–2099. [CrossRef] [PubMed]

6. Gao, J.; Wang, W.; Guo, X.; Zhu, S.; Chen, S.; Zhang, R. Nutrient removal capability and growth characteristics of Iris sibirica in subsurface vertical flow constructed wetlands in winter. *Ecol. Eng.* **2014**, *70*, 351–361. [CrossRef]

7. Jácome, J.A.; Molina, J.; Suárez, J.; Mosqueira, G.; Torres, D. Performance of constructed wetland applied for domestic wastewater treatment: Case study at boimorto (Galicia, Spain). *Ecol. Eng.* **2016**, *95*, 324–329. [CrossRef]

8. Bohórquez, E.; Paredes, D.; Arias, C.A. Vertical flow-constructed wetlands for domestic wastewater treatment under tropical conditions: Effect of different design and operational parameters. *Environ. Technol.* **2017**, *38*, 199–208. [CrossRef] [PubMed]

9. Wu, H.; Fan, J.; Zhang, J.; Ngo, H.H.; Guo, W.; Hu, Z.; Liang, S. Decentralized domestic wastewater treatment using intermittently aerated vertical flow constructed wetlands: Impact of influent strengths. *Bioresour. Technol.* **2015**, *176*, 163–168. [CrossRef] [PubMed]

10. Xu, D.; Li, Y.; Howard, A.; Guan, Y. Effect of earthworm *Eisenia fetida* and wetland plants on nitrification and denitrification potentials in vertical flow constructed wetland. *Chemosphere* **2013**, *92*, 201–206. [CrossRef] [PubMed]

11. Fu, G.; Huang, S.L.; Guo, Z.; Zhou, Q.; Wu, Z. Effect of plant-based carbon sources on denitrifying microorganisms in a vertical flow constructed wetland. *Bioresour. Technol.* **2017**, *224*, 214–221. [CrossRef] [PubMed]

12. Huett, D.O.; Morris, S.G.; Smith, G.; Hunt, N. Nitrogen and phosphorus removal from plant nursery runoff in vegetated and unvegetated subsurface flow wetlands. *Water Res.* **2005**, *39*, 3259–3272. [CrossRef] [PubMed]

13. Shen, Z.; Zhou, Y.; Liu, J.; Xiao, Y.; Cao, R.; Wu, F. Enhanced removal of nitrate using starch/PCL blends as solid carbon source in a constructed wetland. *Bioresour. Technol.* **2015**, *175*, 239–244. [CrossRef] [PubMed]

14. Li, P.; Zuo, J.; Xing, W.; Tang, L.; Ye, X.; Li, Z. Starch/polyvinyl alcohol blended materials used as solid carbon source for tertiary denitrification of secondary effluent. *Environ. Sci.* **2013**, *25*, 1972–1979. [CrossRef]

15. Wen, Y.; Chen, Y.; Zheng, N.; Yang, D.H.; Zhou, Q. Effects of plant biomass on nitrate removal and transformation of carbon sources in subsurface-flow constructed wetlands. *Bioresour. Technol.* **2010**, *101*, 7286–7292. [CrossRef] [PubMed]

16. Shao, L.; Xu, Z.X.; Jin, W.; Yin, H.L. Rice husk as carbon source and biofilm carrier for water denitrification. *Pol. J. Environ. Stud.* **2009**, *18*, 693–699. [CrossRef]

17. Chen, Y.; Wen, Y.; Zhou, Q.; Vymazal, J. Effects of plant biomass on nitrogen transformation in subsurface-batch constructed wetlands: A stable isotope and mass balance assessment. *Water Res.* **2014**, *63*, 158–167. [CrossRef] [PubMed]

18. Yang, X.L.; Jiang, Q.; Song, H.L.; Gu, T.T.; Xia, M.Q. Selection and application of agricultural wastes as solid carbon sources and biofilm carriers in MBR. *J. Hazard. Mater.* **2015**, *283*, 186–192. [CrossRef] [PubMed]

19. Xu, Z.X.; Shao, L.; Yin, H.L.; Chu, H.Q.; Yao, Y.J. Biological denitrification using corncobs as a carbon source and biofilm carrier. *Water Environ. Res.* **2009**, *81*, 242–247. [CrossRef] [PubMed]

20. Li, G.; Chen, J.; Yang, T.; Sun, J.; Yu, S. Denitrification with corncob as carbon source and biofilm carriers. *Water Sci. Technol.* **2012**, *65*, 1238–1243. [CrossRef] [PubMed]

21. Yao, C.Y. Study on Adding Carbon Source to Strengthen Denitrification in Artificial Wetland. Master's Thesis, Dongbei University, Shenyang, China, 2014.

22. Dong, H.Z. Simulation and Regulation of Nitrogen and Phosphorus Pollution in Irrigation Area Based on SWAT Model. Master's Thesis, Chinese Academic Agricultural Science, Beijing, China, 2011.

23. Chang, Y.; Wang, T.; Wang, H.; Chu, Z.; Hang, Q.; Liu, K. The long-term nitrogen removal efficiency from agricultural runoff in phragmites Australis packed surface flow constructed wetland. *J. Environ. Eng. Technol.* **2016**, *6*, 453–461.

24. Jun, C.; Lai, C.Y.; Jun, Z.S.; Ping, L.F.; Qun, X.Y. Degradation test of agricultural non-point source pollution in gully and pond wetland. *Water Res. Power* **2012**, *10*, 107–109.

25. Wang, L.L.; Zhao, L.; Tan, X. Influence of different carbon source and ratio of carbon and nitrogen for water denitrification. *Environ. Prot. Sci.* **2004**, *24*, 45–47.

26. Baker, L.A. Design considerations and applications for wetland treatment of high-nitrate waters. *Water Sci. Technol.* **1998**, *38*, 389–395.

27. Chinese State Environmental Protection Administration. *Water and Wastewater Monitoring Methods*, 4th ed.; Chinese Environmental Science Publishing House: Beijing, China, 2002.

28. Chen, J.; Wei, X.D.; Liu, Y.S.; Ying, G.G.; Liu, S.S.; He, L.Y.; Yang, Y.Q. Removal of antibiotics and antibiotic resistance genes from domestic sewage by constructed wetlands: Optimization of wetland substrates and hydraulic loading. *Sci. Total Environ.* **2016**, *565*, 240–248. [CrossRef] [PubMed]

29. Li, X.; Jia, Y.; Li, B.; Du, B.; Gao, L. Research on pretreatment methods and carbon releasing property of constructed wetland plant as slow-releasing carbon source. *Technol. Water Treat.* **2013**, *39*, 40–46.

30. Waksman, S.A.; Stevens, K.R. A system of proximate chemical analysis of plant materials. *Ind. Eng. Chem. Anal. Ed.* **2002**, *2*, 167–173. [CrossRef]

31. Braskerud, B.C. Factors affecting nitrogen retention in small constructed wetlands treating agricultural non-point source pollution. *Ecol. Eng.* **2002**, *18*, 351–370. [CrossRef]

32. Song, A.H.; Shen, Z.Q.; Zhou, Y.X.; Liu, S.; Xiao, Y.; Miao, Y. Research on treating dispersed piggery rinse water using rice straw as solid carbon source. *Zhongguo Huanjing Kexue/China Environ. Sci.* **2015**, *35*, 2052–2058.

33. Fang-Ying, J.I.; Yang, Y.G.; Wan, X.J.; Ying, H.E. Effects of carbon source types on operation of denitrifying phosphorus removal system. *China Water Wastewater* **2010**, *26*, 5–9.

34. Xu, J.H.; He, S.B.; Wu, S.Q.; Huang, J.C.; Zhou, W.L.; Chen, X.C. Effects of HRT and water temperature on nitrogen removal in autotrophic gravel filter. *Chemosphere* **2016**, *147*, 203–209. [CrossRef] [PubMed]

35. Boroomandnasab, S. The effects of substrate type, HRT and reed on the lead removal in horizontal subsurface-flow constructed wetland. *Desalin. Water Treat.* **2015**, *56*, 3357–3367.

36. Luo, P.; Liu, F.; Liu, X.; Wu, X.; Yao, R.; Chen, L.; Wu, J. Phosphorus removal from lagoon-pretreated swine wastewater by pilot-scale surface flow constructed wetlands planted with *Myriophyllum aquaticum*. *Sci. Total Environ.* **2017**, *576*, 490–497. [CrossRef] [PubMed]

37. Pietro, K.C.; Ivanoff, D. Comparison of long-term phosphorus removal performance of two large-scale constructed wetlands in South Florida, USA. *Ecol. Eng.* **2015**, *79*, 143–157. [CrossRef]

water MDPI

Article

Experimental Study on the Potential Use of Bundled Crop Straws as Subsurface Drainage Material in the Newly Reclaimed Coastal Land in Eastern China

Peirong Lu [1,2], Zhanyu Zhang [1,2,*], Genxiang Feng [1,2], Mingyi Huang [1,2] and Xufan Shi [1,2]

1   Key Laboratory of Efficient Irrigation-Drainage and Agricultural Soil-Water Environment in Southern China of Ministry of Education, Hohai University, Nanjing 210098, China; lupeirongaaron@126.com (P.L.); fenggxhhu@126.com (G.F.); 160202060001@hhu.edu.cn (M.H.); shixufan@hhu.edu.cn (X.S.)
2   College of Water Conservancy and Hydropower Engineering, Hohai University, Nanjing 210098, China
*   Correspondence: zhanyu@hhu.edu.cn; Tel.: +86-25-8378-6947

Received: 29 November 2017; Accepted: 29 December 2017; Published: 2 January 2018

Abstract: Initial land reclamation of the saline soils often requires higher drainage intensity for quick leaching of salts from the soil profile; however, drainage pipes placed at closer spacing may result in higher cost. Seeking an inexpensive degradable organic subsurface drainage material may satisfy such needs of initial drainage, low investment and a heathy soil environment. Crop straws are porous organic materials that have certain strength and endurance. In this research, we explored the potential of using bundled maize stalks and rice straws as subsurface drainage material in place of plastic pipes. Through an experimental study in large lysimeters that were filled with saline coastal soil and planted with maize, we examined the drainage performance of the two organic materials by comparing with the conventional plastic drainage pipes; soil moisture distribution, soil salinity changed with depth, and the crop information were monitored in the lysimeters during the maize growing period. The results showed that maize stalk drainage and the rice straw drainage were significantly ($p < 0.05$) more efficient in removing salt and water from the crop root zone than the plastic drainage pipes; they excelled in drainage rate, leaching fraction, and lowering water table; and their efficient drainage processes lowered salt stress in the crop root zone and resulted in a slightly higher level of biomass. The experimental results suggest that crop straws may be used as a good organic substitute for the plastic drainage pipes in the initial stage land reclamation of the saline coastal soils.

Keywords: subsurface drainage; soil salinity; salt leaching; maize stalk; rice straw

1. Introduction

Development of the coastal mudflat area for agricultural use has been a continuing effort in eastern China. It has been an important regional practice to offset the negative impact of fast population growth and urbanization progress on farmland shortage [1–3]. However, the newly reclaimed coastal mudflat areas generally have brackish shallow groundwater table and high content of salt in the soils, which impede the growth of plants [4]. To make the soils suitable for crop production, land drainage is required to lower the water table and leach the soluble salts from the soil profile. Subsurface drainage with perforated plastic pipes has been a common practice worldwide to control groundwater table or to remove salts from soil profile through leaching irrigation and drainage [5–7]. By lowering water table, subsurface drainage also improves soil aeration at sub-layers and promotes water infiltration, leading to improved development of crop roots and higher crop yields [8,9].

The eastern coastal area of China generally has a humid climatic condition with annual rainfall above 800 mm [10]. Salt leaching in humid area can be accomplished by rainfall when proper drainage system is in place. In the initial stage of reclaiming the saline soils, however, higher drainage intensity, or closely spaced drainage pipes is often required to speed up the salt leaching, or the soil remediation process [11–14]. However, as the soil salinity decreases with the cultivation and the drainage leaching processes, drainage intensity should be lowered to encourage more rainfall storage in soil for improved water use efficiency [15] and prevent losses of soil nutrients from excessive leaching [16,17]. That is, when the soil salinity is lowered to a safe level, high intensity drainage becomes unnecessary. Additionally, the installed underground pipes may become redundant and wasteful considering the relatively high initial cost of subsurface drainage system construction. Therefore, seeking for an inexpensive material that can automatically degraded over time to replace the traditional subsurface plastic drainage facilities for initial land reclamation use may lower the cost for agricultural development and be more environmental friendly in the coastal region.

As byproducts of crop production, straws are traditionally burned after harvest to clear the limited crop fields more quickly with little cost. The smoke emissions from straw burning have been reported as a cause of air pollution in many developing countries [18–20]. It has been blamed as one of the causes for the heavy smog in China during the harvest season [21,22]. To explore their potentials uses, crop straws have been studied in many aspects to discover their applications, such as surface mulching to increase soil temperature and retard the loss of moisture from the root zone [23–25]. Crop straws/stalks are important organic fertility resources for soils [26]; straw interlayer, i.e., straw or mixture of straw and soil buried at different soil depth, has been studied as an agronomic measure to increase moisture storage and organic matter contents, and reduce soluble salt concentrations in the soil during the growing season [27].

Because crop straws are porous organic materials that have certain strength and endurance [28–30], they may be used as a temporary subsurface drainage material when buried underneath crop fields. For the above mentioned land reclamation of the saline coastal soils, crop straws might be an ideal candidate for the initial stage subsurface drainage material. The natural degradation of the organic material in later stage is desired when the soil salinity is under control through the leaching process. Existing research on the crop residue reuses mainly focused on using straws or stalks as isolation layers to reduce soil water loss or to avoid soil salinity buildup; little attention has been paid to the potential use of crop straw as an organic alternative for short term subsurface drainage material.

In this study, we explored the potential of using crop straw (maize stalk and rice straw) as subsurface drainage material. Our hypothesis was that the bundled crop straw buried underground may act like drainage pipes that remove soil water and the dissolved salts in the saline coastal soils in the initial land reclamation stage; their gradual decomposition would be desired, as the soils become salt free later and the field drainage intensity needs to be reduced. The decomposed crop straw may become soil amendments left in the crop fields. As the first step of the research, in this paper, we examined the drainage performance of the maize stalks by comparing it with the conventional corrugated plastic pipes through a lysimeter experiment. The specific objectives of this study were to:

(1) Compare the soil moisture distributions in saline soils as affected by subsurface drainage with bundled maize stalks, rice straws and perforated plastic pipes;
(2) Examine drainage effect of the bundled maize stalks and rice straws in lowering water table and discharge drainage water as compared to the perforated plastic pipes; and
(3) Compare the effects of the three drainage materials on maize yield and root growth to confirm applicability of the straw drainage in early stage of land reclamation of the saline coastal soils.

2. Materials and Methods

2.1. Site Description and Experimental Setup

The experimental study was conducted in 2016 at the testing ground of the Key Laboratory of Efficient Irrigation-Drainage and Agricultural Soil-Water Environment in South China, Ministry of Education in Nanjing, China (118°60′ E, 31°86′ N). The area has a mean annual temperature of 15.3 °C, and a subtropical monsoon climate with hot, wet summers and dry, windy winters [31]. An automated weather station at the experimental site recorded the daily air temperature, precipitation, wind speed, relative humidity, and solar radiation during the study period. Twelve lysimeters of 2.5 m × 2 m × 2 m (length × width × depth) were employed for the experiment. Figure 1 presents a sketch of the lysimeter system. As displayed in Figure 1c, there are multiple drainage outlets through the concrete wall of each lysimeter, subsurface drainage discharges were collected through these outlets using plastic measuring buckets (Figure 1g) in the underground gallery. The water table in each lysimeter can be observed in manometers (Figure 1b) that are attached to the lysimeter wall underground.

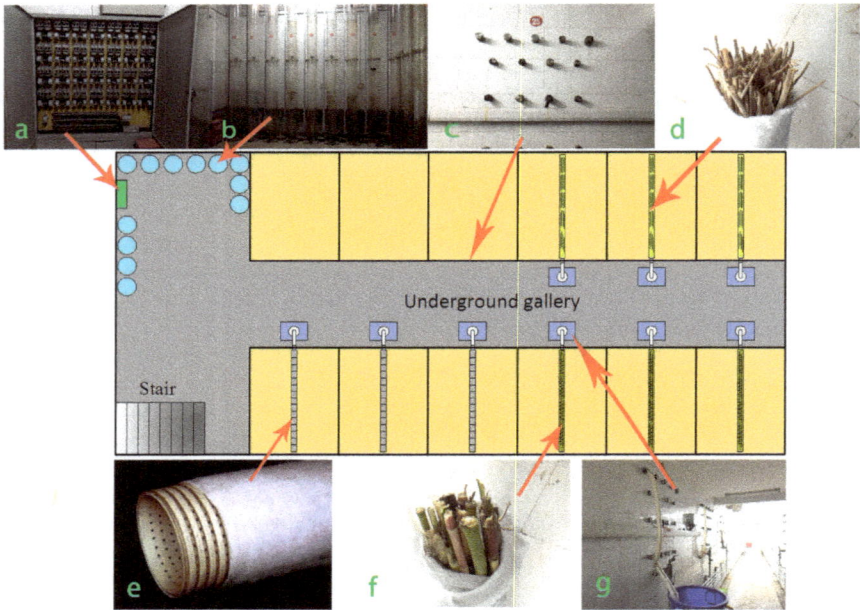

Figure 1. The experimental setup of the 12 drainage lysimeters (including three replicates for each treatment). The components are: (**a**) the regulating switch; (**b**) manometers; (**c**) drainage outlets; (**d**) the rice straw bundle; (**e**) the perforated plastic pipe; (**f**) the maize stalk bundle; and (**g**) the drain flow collection bucket.

During the experiment, the groundwater level was set at 1.2 m from the soil surface, which is based on the average condition of water table depth in the eastern coastal areas of China [32]. Figure 2 shows that the water table in each lysimeter was maintained using a Marriotte bottle system, and an electronic sensor detects water table fall and turns on the water pump automatically to replenish water to the lysimeter. When the water table depth in each lysimeter falls below 1.2 m as result of the crop evapotranspiration, the electronic sensor would turn on the pump to raise the water table up to 1.2 m from the soil surface and the amount of groundwater supply to the plots was recorded using a water meter.

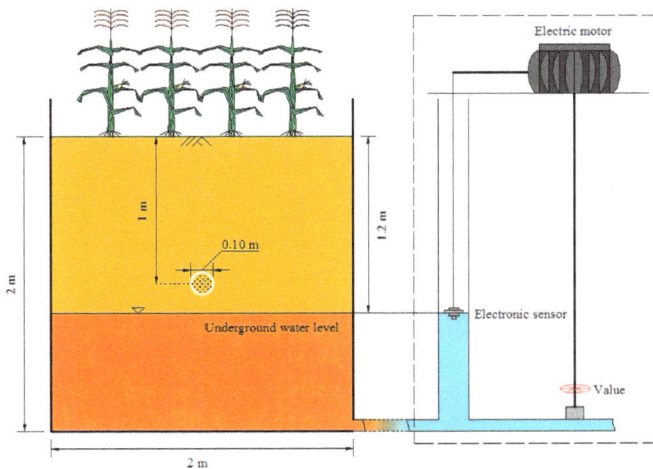

Figure 2. Sketch of the experimental lysimeter profile.

The lysimeters were filled with soil excavated from a newly reclaimed land area in eastern coast of China (Dongtai City, Jiangsu Province, China). The soil was air dried and passed through a 5-mm sieve before filling into the lysimeters. The soil particle analysis showed that the lysimeter soil consists of 22.47% clay (0–0.002 mm), 36.19% silt (0.002–0.02 mm) and 41.34% sand (0.02–2 mm). The soil can be classified as loam based on the USDA soil texture triangle. The measured average soil salinity ($EC_{1:5}$) was 2.07 dS/m and the soil pH was 8.11. The measured soil porosity was 47.54% and the bulk density was 1.34 g·cm^{-3}. The measured field capacity of the 0–60 cm root zone soil was 33%.

Perforated corrugated high-density polyethylene pipes were chosen as the conventional drainage facilities. The pipes were 10 cm in diameter and 180 cm in length, wrapped with nonwoven fabrics. Rice straws and maize stalks were both bundled into conduits of 10 cm in diameter and 180 cm long using plastic ties and enveloped with the same non-woven fabrics. As shown in Figures 1 and 2, all drainage pipes were laid in the middle of each lysimeter along the length at depth of 1.0 m, making the drain spacing as 2 m each.

Maize (*Zea mays* L.) is one of the most widely grown crops in the world. It has been classified as a drought-tolerant crop that is moderately sensitive to soil salinity [33]. Existing studies on maize cultivation showed that water use efficiency, yield and root growth of maize are negatively affected by the soil salinity [34,35]. The maize cultivar (Suyu 29) was sowed on 1 July and harvest on 21 October 2016. Each lysimeter had 24 maize plants with row spacings of 0.40 m and 0.38 m. Maize commonly requires 300–500 mm water during its life cycle under the climate conditions in southeastern China [36]. During the experiment, four fresh water (EC = 0.52 ms/cm) irrigations of 40 mm, 80 mm, 120 mm and 160 mm water were applied separately on 20 July, 16 August, 8 September and 30 September. Irrigation to the lysimeters was controlled by an electromagnetic valve and the amount of irrigation were recorded via a water meter. During the experiment, rainfall effect was excluded by a large automatic rain-shelter that kept all lysimeters free from rainfall. Consequently, the water supply to maize growth was from four irrigation events and groundwater contribution during the entire growing season.

There are four drainage treatments in the experiment, and each treatment was replicated three times, i.e., three lysimeters without subsurface drainage were kept as the control (CK), three lysimeters were installed with the conventional high-density polyethylene plastic drainage pipes (HPD), three lysimeters were installed with the bundled rice straw drainage (RSD) modules and three were installed with the bundled maize stalk drainage (MSD) modules. The irrigation and drainage conditions of all treatments were kept the same.

2.2. Sample Collection and Analysis

To reveal the differences in soil water distribution between the irrigation intervals (around 20 days), the soil water content was monitored 2–3 days after irrigation (i.e., 22 July, 19 August, 10 September and 2 October) and one day before next irrigation or harvest (15 August, 7 September, 29 September and 20 October). Using a soil auger with diameter of 1.2 cm, soil samples were collected at depths of 0–20, 20–40, 40–60 and 60–80 cm at three random locations in each lysimeter; all samples were oven dried at 105 °C to a constant weight to calculate their gravimetric water content (%).

The soil samples for salinity test were taken with the soil auger (40 mm diameter, 90 cm long) at the same layers as the soil moisture tests. The sampling was scheduled for five main growth stages, i.e., seeding stage (19 July), jointing stage (10 August), tasseling stage (31 August), filling stage (12 September) and full ripe stage (29 September). Soil samples were air dried and sieved through 0.5 mm screen; they were then wetted with fresh water (EC of 5~7 μs/cm) before measuring the electrical conductivity of 1:5 soil–water leachate ($EC_{1:5}$) using the DDBJ-350 EC meter (Shanghai INESA Scientific Instrument Co., Ltd., Shanghai, China). All sampling holes in the lysimeters were filled with the surrounding soil after each sample collection.

The groundwater table depth (m) was recorded on an hourly basis during each irrigation event, and the drainage rate (mm/h) was measured one hour after flow started. Water samples were collected with plastic measuring bucket from each treatment during the drainage process, and the drainage water salinity was measured with the DDBJ-350 EC meter.

Upon harvest, three maize plants were harvested randomly for measuring the aboveground biomass and the grain yield in each lysimeter. Plant materials were oven dried at 75 °C to the constant weight. Plant height, stem diameter, dry shoot weight, 100-grain weight and grain yield were measured accordingly. For root length and dry weight measurement, undisturbed soil samples were collected by carefully digging cubic blocks (20 cm × 20 cm × 20 cm) centering a maize plant roots; four soil cubic samples were extracted from one plant at 20 cm depth interval down to 80 cm. After sampling the attached soil was washed out and roots were sieved through a 1 mm screen filter. The clean roots were stored in refrigerator before measurement for root length. A high definition scanner (Epson Perfection V700) was used to generate image files of roots, and WinRHIZO (Regent Instruments Inc., Quebec City, QC, Canada) was used to measure total root length we obtained from each soil cubic core. Roots were then recovered and dried at 75 °C until showing constant weight. The root length density (RLD) (cm/cm^3) was calculated by dividing the total root length (cm) with the volume (cm^3) of the sampling core, and the root weight density (RWD) (mg/cm^3) was calculated by dividing the total root dry weight (g) with the volume (cm^3) of the sampling core.

2.3. Evaluation Methods

The soil moisture variation rate (MSV) during each irrigation interval was calculated by the following equation:

$$\text{MSV} = \frac{\theta_0 - \theta_1}{\theta_0} \tag{1}$$

where θ_0 is the soil gravimetric water content measured 2–3 days after irrigation events, and θ_1 is the soil water content measured one day before next irrigation event or harvest.

The soil desalination rate was calculated with the following equation:

$$S_d = \frac{C_0 - C}{C_0} \times 100\% \tag{2}$$

where C_0 is the initial electrical conductivity ($EC_{1:5}$) in ms/m, and c is the electrical conductivity ($EC_{1:5}$) after crop harvest in ms/m.

The efficiency of subsurface drainage in leaching salt from the soil profile is evaluated with the leaching fraction (LF) calculated with the following equation:

$$LF = \frac{V_d}{V_i} \times 100\% \tag{3}$$

where V_d is the subsurface drainage discharge collected in the plastic bucket after each irrigation event of certain volume, V_i.

2.4. Statistical Analysis

Significant differences in different soil moisture and soil desalination rate were analyzed using one-way ANOVA, and the differences were considered statistically significant at $p < 0.05$. Duncan's multiple-range test was used for comparisons of means at the 0.05 level of significance ($n = 3$) in terms of plant height, stem diameter, dry shoot weight, 100-grain weight and grain yield. SPSS 20.0 was used for statistical analyses with the collected data (SPSS, Chicago, IL, USA).

3. Results and Discussion

3.1. Variability of Soil Moisture Content

As shown in Figure 3, the soil moisture distribution of the straw drainage treatments generally had a similar trend as that in the plastic pipe drainage treatments under same irrigation management; the average soil moisture along the profile decreased slightly in the following sequence: CK > MSD > RSD > HPD. The soil moisture level in the top 30 cm in the drained plots dropped more quickly than the control plots. Soil water contents all increased with the depth of soil layers or the increased irrigation amount. The soil moisture profiles varied less obviously in the third and the fourth irrigation intervals due to the reduced water requirement of maize in the late growth stages and lower daily evaporation in September and October. Similar results of drainage effects on soil water conservation distribution have been reported by Feng et al. [37] and Chang et al. [38].

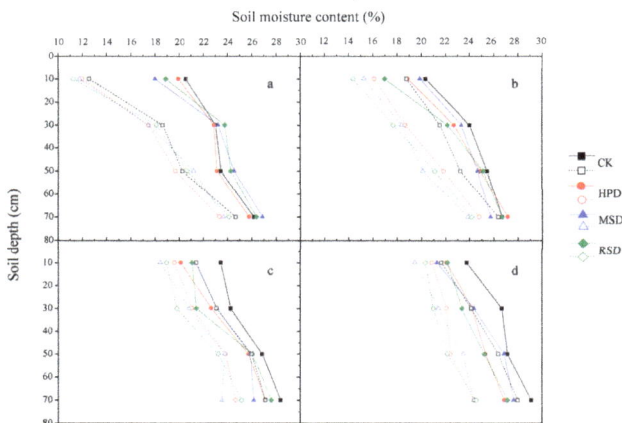

Figure 3. Measured soil water content along depth in different treatment plots during the: (**a**) first irrigation interval; (**b**) second irrigation interval; (**c**) third irrigation interval; and (**d**) fourth irrigation interval. Solid lines represent the water content 2–3 days after irrigation and dash lines represent the water content tested one day before next irrigation event or harvest. (CK is the control plot without subsurface drainage, HPD is the treatment plot with perforated plastic pipes, MSD is the treatment plot with maize stalks, and RSD is the plot with rice straw.)

The variations in soil water content were analyzed for each treatment based on the index of soil moisture variation rate (MSV) for four evaporation periods, during which there were no irrigation or drainage, soil moisture changes were due to evapotranspiration process only. Table 1 lists MSV in different evaporation period and the ANOVA results for the drainage treatments. Considering the low water demand of maize in the seedling stage, the first irrigation application was designed as 40 mm only, which produced no drainage flow in all treatments, so there was no significant ($p > 0.05$) difference among the four treatments. The MSVs for MSD and RSD at the soil layers of 0–20 cm and 20–40 cm were generally higher than that for HPD, and the differences were significant ($p < 0.05$) in the later three periods. However, in the deeper soil layers (40–80 cm), the only significant difference was found at the 40–60 cm layer in the second evaporation period. This showed that subsurface drainage via crop straws removed more soil water than plastic drainage pipes, displaying better soil drainage performance under the irrigation application, especially in the root zone layer (0–40 cm).

Table 1. The average soil moisture variation rate (%) of treatments in four evaporation stages.

Soil Layer (cm)	Treatments	First Evaporation Period (23 July–15 August)	Second Evaporation Period (18 August–7 September)	Third Evaporation Period (10–29 September)	Fourth Evaporation Period (2–20 October)
0–20 cm	CK	38.91% a	7.76% c	6.61% b	7.92% b
	HPD	39.86% a	14.72% b	8.20% b	8.55% ab
	MSD	34.63% a	23.20% a	9.45% a	10.33% a
	RED	40.55% a	15.32% b	10.10% a	9.26% a
20–40 cm	CK	19.07% a	10.33% b	4.74% b	7.11% b
	HPD	23.73% a	17.64% ab	7.65% a	10.15% ab
	MSD	24.86% a	21.28% a	8.12% a	12.23% a
	RED	23.77% a	20.27% a	8.87% a	11.16% a
40–60 cm	CK	13.65% a	8.76% b	3.28% b	2.80% b
	HPD	14.97% a	12.29% ab	7.27% a	9.65% a
	MSD	13.65% a	18.48% a	8.19% a	12.66% a
	RED	14.76% a	16.12% a	10.89% a	12.33% a
60–80 cm	CK	7.60% b	1.16% b	4.37% b	3.91% b
	HPD	9.51% ab	8.70% a	9.03% a	9.34% a
	MSD	11.95% a	7.23% a	10.09% a	11.86% a
	RED	8.49% ab	9.38% a	10.91% a	9.51% a

Note: Values followed by different letters within a column are significantly different at the significance level of 0.05.

3.2. Drainage Performance

Before each irrigation the depth to water table was maintained at constant depth of 1.2 m in all lysimeters, thus the groundwater control system was operated to keep that level by supply water during each evaporation period (Table 2). The first irrigation of 40 mm produced no subsurface drainage, so our analysis focused on the later three irrigation events.

Table 2. Measured irrigation and drainage volume in all treatments.

Treatment	Irrigation Volume (mm)				Groundwater Supply (mm)				Subsurface Drainage (mm)			
	20 July	16 August	8 September	30 September	23 July–15 August	18 August–7 September	10–29 September	2–20 October	20–21 July	16–18 August	8–9 September	30 September–1 October
CK	40	80	120	160	100.66	12.42	7.86	5.96	0	0	0	0
HPD	40	80	120	160	96.19	52.91	33.32	25.83	0	26.69	45.79	74.83
MSD	40	80	120	160	105.28	57.47	36.17	28.04	0	29.42	57.14	93.86
RSD	40	80	120	160	103.11	56.85	35.78	27.73	0	28.21	51.89	84.03

3.2.1. Variation of Water Table Depth and Drainage Rate

Figure 4 shows that the water table rose quickly after irrigation and lowered gradually due to subsurface drainage effect; however, the drainage flow hydrographs showed a relatively symmetrical shape from rise to fall, even though the irrigation amounts were larger in the third and fourth irrigation event. The drainage rate and water table rise among treatments were in sequence of MSD > RSD > HPD for the three different drainage pipes. Regardless of irrigation amount, treatments with HPD, RSD and MSD started to drain when the water table was up to 1.01, 1.11 and 1.06 m, respectively. The crop straw treatment plots started drainage earlier than that of plastic pipes, but the difference in the starting time among different treatments were narrowed by the increasing irrigation amount. Drainage duration (from the beginning to the end of the drainage process) of treatments with MSD and RSD were shorter than those with HPD. Under all treatments, the highest groundwater level was observed at the time when the drainage rate reached the peak value. Liu et al. [39] reported that lowering groundwater level can temporarily increase the rate of infiltration. However, in Figure 4, the peak of water table rise for HPD was delayed comparing with MSD and RSD; the peak values were in the order of HPD > RSD > MSD. The drainage rate in the MSD and RSD plots increased faster than that in the HPD plot before reaching the maximum. This is mainly due to the higher permeability of crop straws, which allowed more soil water flow into the drainage module.

Shallow water table or prolonged periods of waterlogging may cause decreased crop production [40]. Figure 4 shows that subsurface drainage in the MSD and RSD plots occurred more rapidly, in approximately one-third to one-fourth of time of that in HPD. Thus, we can conclude that the higher drainage rate in crop straw treatments may lower water table more quickly after heavy rainfall or large leaching irrigation.

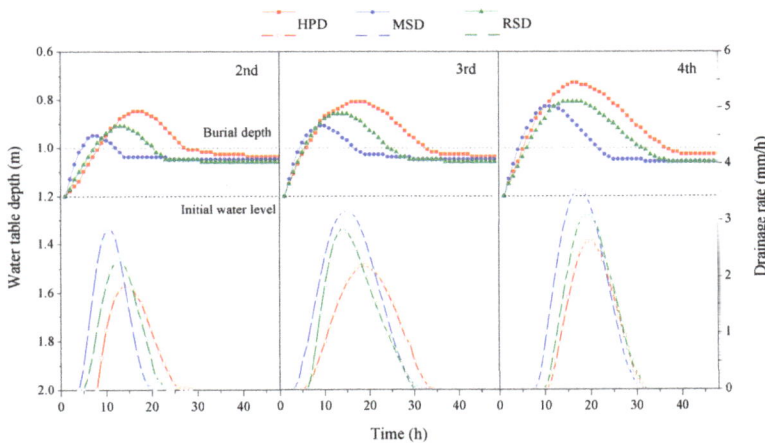

Figure 4. Measured water table fluctuation and subsurface drainage hydrograph from different treatments after three irrigation events (lines with solid markers represent the water table depth, and lines with hollow markers represent the subsurface drainage rate).

3.2.2. Drainage Effect

Figure 5 lists the average drainage rate (ADR), leaching fraction (LF), drainage water salinity (DWS) of different treatments, and the ANOVA results. The ADR and LF were positively correlated with the increasing amount of irrigation water, while the DWS was inversely correlated with the amount of irrigation. The ADR and LF of HPD were significantly lower than that of MSD and RSD in terms of the magnitude of drainage events regardless of the irrigation volume (Figure 5a,b). Higher ADR and LF indicate that the drainage system can discharge more water through the subsurface drains, and consequently leach more salts from the soil profile [41,42]. The higher value in ADR and LF with straw drainage may due to the beneficial effects of the multiple flow paths along the bundled maize stalks or rice straws, resulting in improved soil water transmission along the straw drainage line [43,44]. Although HPD had higher value of DWS than that from MSD and RSD after the second irrigation and the fourth irrigation (Figure 5c), the measured average DWS were 9.21 ds/m for HPD, 9.11 ds/m for MSD and 9.06 ds/m for RSD, showing insignificant ($p > 0.05$) difference among the three drainage treatments.

The above results proved that rice straws and maize stalks can be considered as potential subsurface drainage materials for targeting water table control. They presented similar or better performances to the traditional plastic pipes in salt leaching.

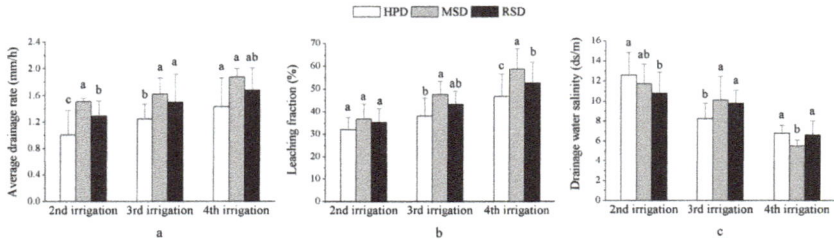

Figure 5. Drainage indices of all treatments during the different irrigation events: (**a**) average drainage rate; (**b**) leaching fraction; and (**c**) drainage water salinity. Different lowercase letters indicate significant ($p < 0.05$) differences among the treatments of the same irrigation event.

3.3. Variations of Soil Salinity

Table 3 lists the ANOVA results for the measured soil salinity ($EC_{1:5}$) from different lysimeters during the five main growth stages of maize during the experiment. The experimental lysimeters were close to each other and they all had similar soil salinity. The measured average soil salinity from all soil samples taken on 19 July 2016 was used as the initial soil salinity. High salinity of soil may lead to salt accumulations on the soil surface if no proper drainage is available for timely discharge of the excess water in the soil profile [45], which is why the soil salinity increased on 10 August, especially at the upper layer (0–40 cm), in all treatment plots. With subsurface drainage, large amounts of salt were removed from the soil profile through the subsurface drainage system in HPD, MSD and RSD treatments. Based on the observations, soil layers at 0–20 cm and 20–40 cm showed significantly lower S_d in MSD and RSD than HPD (Table 3). The low level soil water transfer in MSD limited upward flux and reduced salt movement to the upper layer of soil [46]. The S_d in the treatments of MSD and RSD were 37.04% and 38.90% in the 40–60 cm layer, and 31.86% and 33.21% in the 60–80 cm layer, respectively. These values were not significantly higher ($p > 0.05$) than those with the HPD treatment in the same soil layers (i.e., 39.16% in the 40–60 cm layer and 34.32% in the 60–80 cm layer). These results suggest that the crop straw drainage modules had better desalination performances in the root zone (0–40 cm), and similar desalinization rate (S_d) was observed in the deeper soil layer (40–80 cm) as compared with the HPD.

In general, the soil salinity significantly decreased with the increasing amount of irrigation water, especially in the last two irrigation events (120 mm on 8 September and 160 mm on 25 September) for HPD and the last three irrigations (80 mm on 16 August, 120 mm on 8 September and 160 mm on 25 September) for MSD and RSD. These results indicate that crop straw drainage may produce a relatively higher soil desalination rate with less irrigation water as compared to the traditional plastic drainage pipes. This may attribute to the reason of greater water penetration through the full range of the crop stalks, while the plastic pipes allowed only partial infiltration [47]. In addition, decreasing soil salinity was observed in MSD as the soil depth increased from 0 to 80 cm. This trend was also observed in RSD but less evident to that in MSD, indicating that the maize stalk drainage modules were the most effective in salt leaching due to their improved drainage through the porous stalks that facilitated faster water movement in all directions, leading to faster soil salinity reduction following the infiltration or evaporation process [48].

Table 3. Measured soil $EC_{1:5}$ (ms/m) and salt accumulation with time and soil depth as influenced by the drainage material.

Soil Depth (cm)	Drainage Type	Seeding Stage	Jointing Stage		Tasseling Stage		Filling Stage		Full Ripe Stage	Soil Desalination Rate
0–20	CK	193.23(11.23)	204.67(15.26)		199.26(11.85)		186.09(8.91)		175.39(12.43)	9.12 c
	HPD	180.32(12.11)	188.71(17.98)		173.29(9.21)		160.82(10.33)		137.22(7.09)	23.90 b
	MSD	188.05(14.32)	201.69(12.13)		165.52(11.92)		141.65(7.38)		125.34(11.45)	33.35 a
	RSD	194.2(9.90)	211.32(14.71)		176.42(14.23)		154.83(12.58)		138.68(13.87)	28.59 ab
20–40	CK	146.77(12.19)	152.52(8.76)	First irrigation (40 mm 20 July)	131.34(12.31)	Second irrigation (80 mm 16 August)	122.85(14.50)	Third irrigation (120 mm 8 September)	119.93(10.54)	18.29 c
	HPD	140.21(11.36)	135.52(12.24)		121.62(10.69)		108.26(10.03)		93.97(5.66)	32.98 b
	MSD	132.08(15.92)	129.46(13.01)		113.79(5.78)		99.78(9.72)		82.11(8.72)	37.83 a
	RSD	138.76(12.36)	133.34(8.98)		121.62(13.26)		100.74(7.15)		85.46(10.20)	38.41 a
40–60	CK	137.6(11.56)	134.9(13.87)		129.4(12.14)		119.02(7.99)	Fourth irrigation (160 mm 25 September)	108.26(8.53)	21.32 b
	HPD	128.88(7.63)	120.11(8.92)		111.69(10.15)		103.13(6.89)		78.41(6.34)	39.16 a
	MSD	126.59(11.33)	118.98(11.64)		109.75(12.40)		98.21(9.41)		79.7(7.56)	37.04 a
	RSD	134.9(10.49)	139.23(8.56)		121.2(10.67)		105.6(8.88)		82.43(6.67)	38.90 a
60–80	CK	129.56(7.81)	119.43(12.59)		125.64(13.41)		119.62(13.44)		107.35(14.10)	17.14 b
	HPD	117.79(9.68)	109.06(10.98)		103.46(12.57)		89.63(7.90)		77.36(10.34)	34.32 a
	MSD	111.95(8.18)	110.89(9.26)		101.3(11.14)		91.44(13.90)		76.28(6.22)	31.86 a
	RSD	121.86(10.79)	114.45(13.12)		107.89(7.63)		100.3(9.59)		81.5(10.59)	33.12 a

Note: Values in parentheses are the standard deviations. Different letters in the row indicate significant differences at a certain soil layer ($p \leq 0.05$).

3.4. Responses of Maize Growth

As displayed in Figure 6, maize grown in the lysimeters with subsurface drainage had greater root growth as compared to that in CK. This is consistent with the research findings that subsurface drainage affected root length density (RLD) along the soil depth [49,50]. In Figure 6a, RLD reached the maximum in the top 20 cm and the minimum in 40–60 cm in all treatments. Close values ($p > 0.05$) of RLD in HPD, MSD and RSD were found in the 0–40 cm soil layer; significant differences were observed in the 60–80 cm soil layer; the RLD values decreased in the sequence of MSD > RSD > HPD. Owing to the less distribution of root in the 40–80 cm layer, no significant ($p < 0.05$) differences were observed in the average root length density among the three treatments.

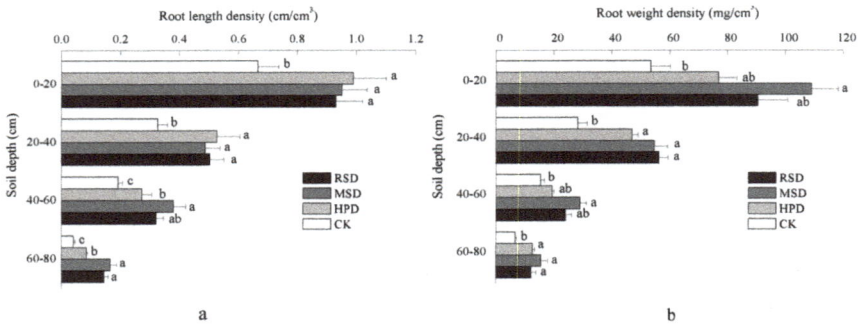

Figure 6. The average individual root length density (**a**); and dry root weight density (**b**) of different treatments in soil cores sampled in different depth. Different lowercase letters indicate significant ($p < 0.05$) differences among the treatments of the same sampling depth.

Corresponding to the variation of RLD, the value of root weight density (RWD) present higher under the MSD and RSD than that with the HPD, particularly in the upper layer. However, under the same planting condition, significantly heavier weight ($p < 0.05$) was found in the average value of RWD among three treatments and they were ranked as MSD (52.12 mg/cm^3) > RSD (45.81 mg/cm^3) > HPD (39.03 mg/cm^3), indicating that crop straws had a positive effect on root matter growth. The better root development of the growing maize inevitably led to improved soil water holding capacity [51], and the improved soil condition encourages more root growth. This may be attributed to the good subsurface drainage that significantly improved the soil properties such as the drainable porosity [52] and the hydraulic conductivity [53] in the soil layers where roots were more concentrated.

In the present study, salt stress from the saline soil affected the maize growth (Table 4). Reduction in plant growth as result of the salt stress has been reported in many studies [54,55]. Plant height, stem diameter, shoot dry weight, 100-grain weight and grain yield were found significantly higher ($p < 0.05$) in HPD (24.60%, 146.16%, 27.14%, 44.35%, and 30.93%, respectively), MSD (39.68%, 230.77%, 41.64%, and 41.32%, respectively) and RSD (43.65%, 261.54%, 39.70%, and 38.64%, respectively) as compared to the control plot without drainage. The utilization of crop straws as drainage materials produced significant ($p < 0.05$) increase in the above ground biomass, including the plant height, stem diameter and shoot dry matter as compared to the plastic pipes. Among the treatments of HPD, MSD and RSD, however, no significant ($p > 0.05$) difference was observed in the values of 100-grain weight and grain yield. Overall, the growth and yield of maize excelled in the lysimeters with the subsurface drainage system; some observed beneficial effect in above ground biomass may be the result of the crop straws that provided more favorable condition for root growth.

Table 4. The Effect of drainage type and salinity of irrigation water on above ground biomass and the grain of maize.

Treatments	Plant Height (cm)	Stem Diameter (cm)	Shoot Dry Weight (g)	100-Grain Weight (g)	Grain Yield (g/plant)
CK	126 (5.3) c	1.3(0.2) b	86.22(4.63) c	13.53(2.96) b	78.77(12.66) b
HPD	157 (4.2) b	3.2(0.3) b	109.62(8.26) b	19.43(3.04) ab	103.13(5.44) b
MSD	176(7.2) ab	4.3(0.5) a	122.12(7.58) ab	20.31(1.86) ab	111.32(11.73) ab
RSD	181(9.4) a	4.7(0.3) a	129.38(13.09) a	18.90(2.56) a	109.21(11.29) a

Note: Different letters after the data in the same column indicate significant differences among treatments at $p < 0.05$.

4. Conclusions

Through parallel experiment in lysimeters, this study examined the potential of using crop straws as subsurface drainage material in the initial land reclamation of the saline coastal soils. Observations through the maize growing period concluded that:

(1) Subsurface drainage with crop straws had positive effect on the salt and water distribution in the soil profile; less soil moisture variation and higher desalination rate were observed in the root zone drained with the bundled crop straw than that drained with the conventional plastic pipes.

(2) The crop straws drainage modules displayed faster soil water transmission properties, as reflected in the quick water table drawdown, greater average drainage rate and leaching fraction.

(3) The crop straw treatments produced better crop growth in terms of the root distribution, plant length, stem diameter and shoot dry weight of the maize, showing potential benefit of the organic drainage material.

The experimental results suggest that using the crop straws as subsurface drainage material may be a good option in the initial land reclamation of the saline soils. Considering the renewable and biodegradable nature of crop straw resources, the application of crop straw as subsurface drainage materials may achieve a good balance between crop production and drainage system construction. The long-term impact of crop straw as subsurface drainage material on drainage water quality and soil properties under different irrigation schedules will be studied in the future.

Acknowledgments: Funding for this research was partially supported by the Natural Science Foundation of China (No. 51579069), the Fundamental Research Funds for the Central Universities (2017B691X14), the Scientific Research Innovation Projects in Jiangsu general Universities (KYCX17-0437), the Priority Academic Program Development of Jiangsu Higher Education Institutions (YS11001), and the China Postdoctoral Science Foundation funded project (2017M621619).

Author Contributions: All authors made a tangible contribution to this manuscript. Peirong Lu, Zhanyu Zhang and Xufan Shi were in charge of the experimental design and operation. Peirong Lu analyzed the results and wrote the manuscript. Zhanyu Zhang and Genxiang Feng reviewed the manuscript. Mingyi Huang provided much help to proofreading and improving the structure of the manuscript. All authors have approved the content of the submitted manuscript.

Conflicts of Interest: The authors declare no conflict of interest.

References

1. Long, X.H.; Liu, L.P.; Shao, T.Y.; Shao, H.B.; Liu, Z.P. Developing and sustainably utilize the coastal mudflat areas in China. *Sci. Total Environ.* **2016**, *569*, 1077–1086. [CrossRef] [PubMed]

2. Li, J.; Pu, L.; Zhu, M.; Zhang, J.; Li, P.; Dai, X.; Xu, Y.; Liu, L. Evolution of soil properties following reclamation in coastal areas: A review. *Geoderma* **2014**, *226*, 130–139. [CrossRef]

3. Bu, N.-S.; Qu, J.-F.; Li, G.; Zhao, B.; Zhang, R.-J.; Fang, C.-M. Reclamation of coastal salt marshes promoted carbon loss from previously-sequestered soil carbon pool. *Ecol. Eng.* **2015**, *81*, 335–339. [CrossRef]

4. Fernandez, S.; Santin, C.; Marquinez, J.; Alvarez, M.A. Saltmarsh soil evolution after land reclamation in atlantic estuaries (Bay of Biscay, North Coast of Spain). *Geomorphology* **2010**, *114*, 497–507. [CrossRef]

5. Qureshi, A.S.; McCornick, P.G.; Qadir, M.; Aslam, Z. Managing salinity and waterlogging in the Indus Basin of Pakistan. *Agric. Water Manag.* **2008**, *95*, 1–10. [CrossRef]
6. Youngs, E.G.; Leeds-Harrison, P.B. Improving efficiency of desalinization with subsurface drainage. *J. Irrig. Drain. Eng. ASCE* **2000**, *126*, 375–380. [CrossRef]
7. Christen, E.W.; Ayars, J.E.; Hornbuckle, J.W. Subsurface drainage design and management in irrigated areas of Australia. *Irrig. Sci.* **2001**, *21*, 35–43. [CrossRef]
8. Mehnert, E.; Hwang, H.-H.; Johnson, T.M.; Sanford, R.A.; Beaumont, W.C.; Holm, T.R. Denitrification in the shallow ground water of a tile-drained, agricultural watershed. *J. Environ. Qual.* **2007**, *36*, 80–90. [CrossRef] [PubMed]
9. Lenhart, C.; Gordon, B.; Gamble, J.; Current, D.; Ross, N.; Herring, L.; Nieber, J.; Peterson, H. Design and hydrologic performance of a tile drainage treatment wetland in Minnesota, USA. *Water* **2016**, *8*, 549. [CrossRef]
10. Gong, D.Y.; Ho, C.H. Shift in the summer rainfall over the Yangtze River valley in the late 1970s. *Geophys. Res. Lett.* **2002**, *29*, 10. [CrossRef]
11. Shao, X.-H.; Chang, T.-T.; Cai, F.; Wang, Z.-Y.; Huang, M.-Y. Effects of subsurface drainage design on soil desalination in coastal resort of China. *J. Food Agric. Environ.* **2012**, *10*, 935–938.
12. Afruzi, A.; Nazemi, A.H.; Sadraddini, A.A. Steady-state subsurface drainage of ponded fields by rectangular ditch drains. *Irrig. Drain.* **2014**, *63*, 668–681. [CrossRef]
13. Manjunatha, M.V.; Oosterbaan, R.J.; Gupta, S.K.; Rajkumar, H.; Jansen, H. Performance of subsurface drains for reclaiming waterlogged saline lands under rolling topography in Tungabhadra irrigation project in India. *Agric. Water Manag.* **2004**, *69*, 69–82. [CrossRef]
14. Sharma, D.R.; Gupta, S.K. Subsurface drainage for reversing degradation of waterlogged saline lands. *Land Degrad. Dev.* **2006**, *17*, 605–614. [CrossRef]
15. Skaggs, R.W.; Breve, M.A.; Gilliam, J.W. Hydrologic and water quality impacts of agricultural drainage. *Crit. Rev. Environ. Sci. Technol.* **1994**, *24*, 1–32. [CrossRef]
16. Yu, S.; Liu, J.; Eneji, A.E.; Han, L.; Tan, L.; Liu, H. Dynamics of soil water and salinity under subsurface drainage of a coastal area with high groundwater table in spring and rainy season. *Irrig. Drain.* **2016**, *65*, 360–370. [CrossRef]
17. Rozemeijer, J.C.; Visser, A.; Borren, W.; Winegram, M.; van der Velde, Y.; Klein, J.; Broers, H.P. High-frequency monitoring of water fluxes and nutrient loads to assess the effects of controlled drainage on water storage and nutrient transport. *Hydrol. Earth Syst. Sci.* **2016**, *20*, 347–358. [CrossRef]
18. Jain, N.; Bhatia, A.; Pathak, H. Emission of air pollutants from crop residue burning in India. *Aerosol Air Qual. Res.* **2014**, *14*, 422–430. [CrossRef]
19. Yaman, B.; Aydin, Y.M.; Koca, H.; Dasdemir, O.; Kara, M.; Altiok, H.; Dumanoglu, Y.; Bayram, A.; Tolunay, D.; Odabasi, M.; et al. Biogenic volatile organic compound (BVOC) emissions from various endemic tree species in Turkey. *Aerosol Air Qual. Res.* **2015**, *15*, 341–356. [CrossRef]
20. Gadde, B.; Bonnet, S.; Menke, C.; Garivait, S. Air pollutant emissions from rice straw open field burning in India, Thailand and the Philippines. *Environ. Pollut.* **2009**, *157*, 1554–1558. [CrossRef] [PubMed]
21. Shi, T.; Liu, Y.; Zhang, L.; Hao, L.; Gao, Z. Burning in agricultural landscapes: An emerging natural and human issue in China. *Landsc. Ecol.* **2014**, *29*, 1785–1798. [CrossRef]
22. Tian, H.; Zhao, D.; Wang, Y. Emission inventories of atmospheric pollutants discharged from biomass burning in China. *Acta Sci. Circumst.* **2011**, *31*, 349–357.
23. Dahiya, R.; Ingwersen, J.; Streck, T. The effect of mulching and tillage on the water and temperature regimes of a loess soil: Experimental findings and modeling. *Soil Tillage Res.* **2007**, *96*, 52–63. [CrossRef]
24. Ramakrishna, A.; Tam, H.M.; Wani, S.P.; Long, T.D. Effect of mulch on soil temperature, moisture, weed infestation and yield of groundnut in northern Vietnam. *Field Crop. Res.* **2006**, *95*, 115–125. [CrossRef]
25. Huang, Y.L.; Chen, L.D.; Fu, B.J.; Huang, Z.L.; Gong, E. The wheat yields and water-use efficiency in the loess plateau: Straw mulch and irrigation effects. *Agric. Water Manag.* **2005**, *72*, 209–222. [CrossRef]
26. Bi, L.; Zhang, B.; Liu, G.; Li, Z.; Liu, Y.; Ye, C.; Yu, X.; Lai, T.; Zhang, J.; Yin, J.; et al. Long-term effects of organic amendments on the rice yields for double rice cropping systems in subtropical China. *Agric. Ecosyst. Environ.* **2009**, *129*, 534–541. [CrossRef]
27. Zhao, Y.; Pang, H.; Wang, J.; Li, Y.; Li, Y. Depth of stover layer for salt management influences sunflower production in saline soils. *Crop Sci.* **2016**, *56*, 1948–1961. [CrossRef]

28. Dias, D.; Lapa, N.; Bernardo, M.; Godinho, D.; Fonseca, I.; Miranda, M.; Pinto, F.; Lemos, F. Properties of chars from the gasification and pyrolysis of rice waste streams towards their valorisation as adsorbent materials. *Waste Manag.* **2017**, *65*, 186–194. [CrossRef] [PubMed]

29. Mitchell, R.D.J.; Harrison, R.; Russell, K.J.; Webb, J. The effect of crop residue incorporation date on soil inorganic nitrogen, nitrate leaching and nitrogen mineralization. *Biol. Fertil. Soils* **2000**, *32*, 294–301. [CrossRef]

30. Mohan, D.; Pittman, C.U.; Steele, P.H. Pyrolysis of wood/biomass for bio-oil: A critical review. *Energy Fuels* **2006**, *20*, 848–889. [CrossRef]

31. Li, C.X.; Zhang, J.Q.; Fan, D.D.; Deng, B. Holocene regression and the tidal radial sand ridge system formation in the Jiangsu coastal zone, east China. *Mar. Geol.* **2001**, *173*, 97–120. [CrossRef]

32. Sun, J.; Kang, Y.; Wan, S. Effects of an imbedded gravel-sand layer on reclamation of coastal saline soils under drip irrigation and on plant growth. *Agric. Water Manag.* **2013**, *123*, 12–19. [CrossRef]

33. Kang, Y.; Chen, M.; Wan, S. Effects of drip irrigation with saline water on waxy maize (*Zea mays* L. Var. *Ceratina kulesh*) in north China plain. *Agric. Water Manag.* **2010**, *97*, 1303–1309. [CrossRef]

34. Lacerda, C.F.; Sousa, G.G.; Silva, F.L.B.; Guimaraes, F.V.A.; Silva, G.L.; Cavalcante, L.F. Soil salinization and maize and cowpea yield in the crop rotation system using saline waters. *Eng. Agricola* **2011**, *31*, 663–675. [CrossRef]

35. Neidhardt, H.; Norra, S.; Tang, X.; Guo, H.; Stuben, D. Impact of irrigation with high arsenic burdened groundwater on the soil-plant system: Results from a case study in the Inner Mongolia, China. *Environ. Pollut.* **2012**, *163*, 8–13. [CrossRef] [PubMed]

36. Wang, Q.; Huo, Z.; Zhang, L.; Wang, J.; Zhao, Y. Impact of saline water irrigation on water use efficiency and soil salt accumulation for spring maize in arid regions of China. *Agric. Water Manag.* **2016**, *163*, 125–138. [CrossRef]

37. Feng, G.; Zhang, Z.; Wan, C.; Lu, P.; Bakour, A. Effects of saline water irrigation on soil salinity and yield of summer maize (*Zea mays* L.) in subsurface drainage system. *Agric. Water Manag.* **2017**, *193*, 205–213. [CrossRef]

38. Chang, T.; Shao, X.; Ye, H.; Li, W.; Zhang, J.; Zhang, Z. Irrigation scheduling for corn in a coastal saline soil. *Int. J. Agric. Biol. Eng.* **2016**, *9*, 91–99.

39. Liu, C.W.; Cheng, S.W.; Yu, W.S.; Chen, S.K. Water infiltration rate in cracked paddy soil. *Geoderma* **2003**, *117*, 169–181. [CrossRef]

40. Nosetto, M.D.; Jobbagy, E.G.; Jackson, R.B.; Sznaider, G.A. Reciprocal influence of crops and shallow ground water in sandy landscapes of the inland pampas. *Field Crop. Res.* **2009**, *113*, 138–148. [CrossRef]

41. Sharma, D.P.; Tyagi, N.K. On-farm management of saline drainage water in arid and semi-arid regions. *Irrig. Drain.* **2004**, *53*, 87–103. [CrossRef]

42. Singh, A. Decision support for on-farm water management and long-term agricultural sustainability in a semi-arid region of India. *J. Hydrol.* **2010**, *391*, 65–78. [CrossRef]

43. Rasool, R.; Kukal, S.S.; Hira, G.S. Soil organic carbon and physical properties as affected by long-term application of FYM and inorganic fertilizers in maize-wheat system. *Soil Tillage Res.* **2008**, *101*, 31–36. [CrossRef]

44. Tejada, M.; Gonzalez, J.L.; Garcia-Martinez, A.M.; Parrado, J. Effects of different green manures on soil biological properties and maize yield. *Bioresour. Technol.* **2008**, *99*, 1758–1767. [CrossRef] [PubMed]

45. Beltran, J.M. Irrigation with saline water: Benefits and environmental impact. *Agric. Water Manag.* **1999**, *40*, 183–194. [CrossRef]

46. Fu, P.; Hu, S.; Xiang, J.; Sun, L.; Su, S.; Wang, J. Evaluation of the porous structure development of chars from pyrolysis of rice straw: Effects of pyrolysis temperature and heating rate. *J. Anal. Appl. Pyrolysis* **2012**, *98*, 177–183. [CrossRef]

47. Araguees, R.; Teresa Medina, E.; Claveria, I. Effectiveness of inorganic and organic mulching for soil salinity and sodicity control in a grapevine orchard drip-irrigated with moderately saline waters. *Span. J. Agric. Res.* **2014**, *12*, 501–508. [CrossRef]

48. Malash, N.M.; Flowers, T.J.; Ragab, R. Effect of irrigation methods, management and salinity of irrigation water on tomato yield, soil moisture and salinity distribution. *Irrig. Sci.* **2008**, *26*, 313–323. [CrossRef]

49. Fiebig, A.; Dodd, I.C. Inhibition of tomato shoot growth by over-irrigation is linked to nitrogen deficiency and ethylene. *Physiol. Plant.* **2016**, *156*, 70–83. [CrossRef] [PubMed]

50. Mathew, E.K.; Panda, R.K.; Nair, M. Influence of subsurface drainage on crop production and soil quality in a low-lying acid sulphate soil. *Agric. Water Manag.* **2001**, *47*, 191–209. [CrossRef]

51. Da Rocha, M.G.; Faria, L.N.; Casaroli, D.; Van Lier, Q.D.J. Evaluation of a root-soil water extraction model by root systems divided over soil layers with distinct hydraulic properties. *Rev. Bras. Cienc. Solo* **2010**, *34*, 1017–1028.

52. Haws, N.W.; Rao, P.S.C.; Simunek, J.; Poyer, I.C. Single-porosity and dual-porosity modeling of water flow and solute transport in subsurface-drained fields using effective field-scale parameters. *J. Hydrol.* **2005**, *313*, 257–273. [CrossRef]

53. Wang, J.M.; Wang, P.; Qin, Q.; Wang, H.D. The effects of land subsidence and rehabilitation on soil hydraulic properties in a mining area in the loess plateau of China. *Catena* **2017**, *159*, 51–59. [CrossRef]

54. Farooq, M.; Hussain, M.; Wakeel, A.; Siddique, K.H.M. Salt stress in maize: Effects, resistance mechanisms, and management. A review. *Agron. Sustain. Dev.* **2015**, *35*, 461–481. [CrossRef]

55. Katerji, N.; van Hoorn, J.W.; Hamdy, A.; Mastrorilli, M. Salinity effect on crop development and yield, analysis of salt tolerance according to several classification methods. *Agric. Water Manag.* **2003**, *62*, 37–66. [CrossRef]

water

MDPI

Article

Assessment of a Field Tidal Flow Constructed Wetland in Treatment of Swine Wastewater: Life Cycle Approach

Tong Wang [1], Ranbin Liu [2], Emmet Mullan [2] and Yaqian Zhao [2,3,*]

[1] School of Civil Engineering, Chang'an University, Xian 710061, China; wangt@chd.edu.cn
[2] UCD Dooge Centre for Water Resources Research, School of Civil Engineering, University College Dublin, Newstead, Belfield, Dublin D04 K3H4, Ireland; liu.ranbin@ucdconnect.ie (R.L.); kate.omeara@ucdconnect.ie (K.O.); emmet.mullan@ucdconnect.ie (E.M.)
[3] Key Laboratory of Subsurface Hydrology and Ecology in Arid Areas (Ministry of Education), School of Environmental Science and Engineering, Chang'an University, Xian 710054, China
* Correspondence: yaqian.zhao@ucd.ie; Tel.: +353-1-7163215

Received: 7 April 2018; Accepted: 26 April 2018; Published: 28 April 2018

Abstract: The spreading of livestock wastewater onto the grassland poses the inevitable risk of pollutants into the surface water or ground water, causing adverse environmental problems. Although the constructed wetlands (CWs) represent a cost-effective treatment system, they fail to achieve satisfactory total nitrogen (TN) removal performance. Dewatered alum sludge (DAS) based CW with tidal flow operation strategy is set up to intensify the TN removal efficiency by creating alternating aerobic and anoxic conditions, which relies on the water pumps instead of air pumps. In the present study, the environmental performance of a four-stage field tidal flow CW system treating swine wastewater was evaluated based on the life cycle assessment (LCA). The contribution of each process in LCA was clarified and compared whereby the potential improvement was indicated for further application. The results showed that the electricity almost dominated all the environmental impact categories while the water pumps (used for creating tidal flow) were the dominant electricity consumer. Moreover, the mitigation effect of vegetation by uptaking CO_2 was relatively marginal. Overall, compared with conventional CWs, the tidal flow CW brought about more adverse impact to the environment although the tidal flow could achieve better treatment efficiency.

Keywords: livestock wastewater; life cycle assessment; nitrogen removal; tidal flow constructed wetland

---

1. Introduction

The agri-food sector is one of Ireland's most important indigenous manufacturing sectors, accounting for the employment of approximately 167,500 people [1]. In 2015, the agri-food sector in Ireland generated 5.7% of gross value and 9.8% of Ireland's merchandise exports [1]. Behind the economic prosperity, there is a large amount of livestock wastewater generation in Ireland. Although the exact wastewater data are not available, the adverse effect of such a large amount of wastewater production and inappropriate management has emerged and attracted attention. In Ireland, it is common in practice to spread livestock wastewater on nearby grassland after anaerobic stabilization [2]. Indeed, it is a convenient option for the wastewater while the nutrients in the wastewater can help grass growth. However, there are also some harmful components, such as antibiotics and heavy metals etc. Moreover, the leakage and diffusion of these substances into the surface water or groundwater could cause severe pollution. As such, Ireland is labelled as one of the nitrate vulnerable zones by the EU (European Union) [3].

Thus, it is highly desirable to develop a sustainable approach to manage the livestock farming wastewater. Indeed, constructed wetlands (CWs) have been recognized as a popular and low-cost technology to treat livestock farming wastewater [4]. Moreover, the application and installment of CWs are highly flexible, catering to various locations and scales, compared with other treatment technologies. However, regarding the high strength livestock wastewater, conventional CWs, including surface flow CWs and subsurface flow CWs or even the hybrid system of vertical- and horizontal-flow combination, cannot achieve satisfying treatment efficiency especially for TN and total phosphorus (TP). Therefore, in recent years, dewatered alum sludge (DAS) was intensively tested as the main wetland substrate for effective phosphorus (P) removal/immobilization, while "tidal flow" operation strategy was developed and had been demonstrated to be a good input in improving the air supply and thus bringing about a better nitrogen (N) removal [5,6].

Alum sludge refers to the drinking water treatment residual when aluminum sulphate was dosed to flocculate the raw water. The use of DAS cakes as substrate in CWs lies in the Al^{3+} in the sludge to adsorb P in wastewater since Al^{3+} and P have strong adsorption affinity from chemistry point of view. Therefore, the alum sludge-based CWs were developed [5]. "Tidal flow" CWs are a variant of passive CWs owning improved treatment performance and capacity [6]. They are operated in accordance with sequencing batch philosophy with a cycle consisting of fill, contact, drain, and rest period sequentially whereby a tide is generated in the bed matrix. In such a tide regime, the redox status in the bed matrix varies with the saturated/unsaturated conditions corresponding to the contact/rest periods. Herein, the nitrification and denitrification can be promoted in a tide cycle, respectively. As such, tidal flow CW seems an alternative to the aeration-intensified CWs to treat livestock farming wastewater. It has been well demonstrated that alum sludge-based tidal flow CWs enable the system to treat the high strength wastewater [6]. It is noted that, in the tidal flow operation, the electricity fueling the air pumps can be avoided. However, the tides in the tidal CWs rely on water pumps to create aerobic and anoxic conditions. Compared with the intensified CWs by air pumps, the electricity consumption transferred to the water pumps. It is fair to say that attention has been paid to the treatment efficiency, while less attention was placed on the entire environment impact of the tidal flow CW system when more pumps were used to create the "tides". Clearly, regarding the novel alum sludge-based tidal flow CWs, there is a "gap" between good treatment efficiency and the overall environmental impact in the literature. The quantification and verification on the environmental performance of the tidal flow CWs is desirable. This forms the basis of the current study.

Life cycle assessment (LCA) is a standardized and sophisticated tool to "compile and evaluate the inputs, outputs and the potential environmental impacts of a product system through its life cycle" [7]. As adopted in the field of wastewater treatment, LCA allows assessing the environmental sustainability over its complete life cycle. Since LCA was applied in wastewater treatment plants (WWTPs) as early as in 1990s, it has been widely applied in research or practical projects of WWTPs. The key step of LCA is to collect the input and output inventory of the target object in as much detailed as possible. The exact procedure of LCA in WWTPs can refer to several comprehensive reviews [7–9].

To address the environment-related issues of the tidal flow CWs in treating swine wastewater, this paper presents a LCA study aimed at looking into the environmental performance of a field work in an Irish farm. All the related processes and materials were classified and evaluated. This LCA was performed through the analysis of seven environmental categories, pinpointing the process contributing the most, and assessing the sensitivity to the background processes. According to the best knowledge of the authors, this study is the first study to explore the insight into the environmental performance of tidal flow CWs.

2. Materials and Methods

2.1. Target Process

The pilot-scale field CW consisted of four identical reed beds (Figure 1) [10]. Each identical bed/stage of the CWs was constructed using a 1100 L plastic bin (108 cm × 94 cm × 105 cm), while the four stages were connected with submersible pumps placed inside a well (40 cm × 40 cm × 100 cm) set within each bin. Each well also served as the sampling port. Each bin was filled with 10 mm gravel at the bottom up to a depth of 10 cm, covered with 65 cm of DAS cakes as the 'medium' layer, and then 10 cm of 20 mm gravel to serve as the distribution layer. The DAS cakes were collected fresh from a drinking water treatment plant in Southwest Dublin. Common reeds, phragmites australis, were planted on top of each stage at the commencement of the experimental trials. The good growth with lush vegetation was observed after 2 months. The characteristics of the swine wastewater and the performance of this tidal flow CW are summarized in Table 1.

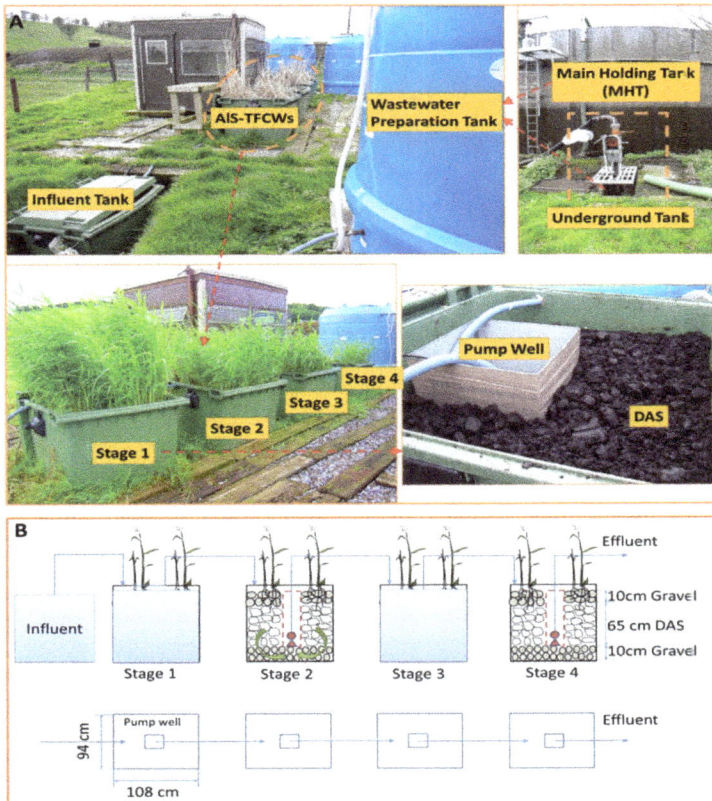

Figure 1. Field photos (**A**) and schematic representation (**B**) of the pilot-scale four stages tidal flow constructed wetlands (CWs).

Table 1. Wastewater characteristics, construction parameters, and performance of the CW [10].

Wastewater (mg/L)					
Composition	Influent	Effluent	Composition	Influent	Effluent
BOD	318	102	PO_4^{3-}-P	20	1.5
COD	446	206	SS	188	68
TN	136	72	Al	0.01	0.07
CW (Each Stage)					
Volume-total	1100 L		DAS	972 kg (75% moisture)	
Volume-working	180 L		Gravel-mass	305 kg	
Cycle	3 cycles/day		Cycle time	8 h	

2.2. Goal and Scope Definition

The targeted product in the present study is a field pilot-scale tidal flow CW established for purifying swine wastewater in a livestock farm. The initial function of the tidal flow CW was to remove pollutants from the influent swine wastewater by the production of purified effluent. Thus, the Functional Unit (FU) of the present assessment is defined in quantitative terms as the production of 1 m^3 of purified wastewater [11]. As the swine wastewater collecting system in the farm had already existed, it was not included in the present LCA. In addition, the construction and demolition phases under the present LCA were excluded.

2.3. Life Cycle Inventory

2.3.1. Input

According to the field work, the main materials and energy input for the construction and operation of the tidal flow CWs include containers, pumps, gravels, DAS, transportation, and electricity. The field tidal flow CW was a newly established system in the farm and five HDPE (high-density polyethylene) plastic bins including four stages and influent tank were used to form the CW system to keep the operation and upgrading flexible. A life time of 10 years was assumed for the replacement of the plastic containers. Each HDPE container was 1100 L in volume and 40 kg in weight. As such, the HDPE needed for manufacturing the containers in terms of one FU could be normalized against the wastewater volume treated in the 10 years (Table 2). Compared with conventional passive CWs, the tidal flow CWs rely on pumps to create the tides. As such, 5 submersible pumps were installed in the system. The rated flow and power are 1800 L/h and 1100 W, respectively. It is hypothesized that they should be renewed every 5 years. Then, the number of pumps per FU can be calculated (Table 2). In terms of the electricity consumption, it mainly came from the pumps. In each cycle, each submersible pump worked for 0.1 h based on the working volume of each unit. Thus, the total electricity input could be calculated. The amounts of gravel and DAS in the systems were listed in Table 1. For DAS, it should be considered to replace to promise the P removal performance when the DAS become fully saturated [12]. To evaluate the worst scenario of the field tidal flow CW, the gravel was assumed to be replaced as well along with the DAS.

For the background processes including the electricity generation, containers/pumps manufacturing, and transportation, the data were derived from the Chinese Life Cycle Database (CLCD) [13]. Electricity from the grid was regarded as the only energy input in the present LCA which comprised of 42%, 25%, 17% and 16% from natural gas, coal, renewable sources, and others [14]. The containers and the pumps were assumed to be transported with a distance of 100 km while the DAS and gravel were assumed to be transported with a distance of 20 km. All the transportation is based on a lorry with a capacity of 16–32 t.

Table 2. LCA inventory of direct input of the tidal CW per FU (1 m^3).

Item	Unit	Value	Item	Unit	Value
COD	g	446	Containers	kg	0.055
TN	g	136	Tran$_{container}$	t·km	0.0055
TP	g	20	Gravel	kg	2.36
Electricity	kWh	2.36	Tran$_{gravel}$	t·km	0.047
Pumps	-	0.00186	DAS	kg	7.55
Tran$_{pump}$	t·km	2.62	Tran$_{alum}$	t·km	0.151
Vegetation	m^2	0.011	-	-	-

2.3.2. Emissions

All the processes, i.e., electricity and chemicals production, wastewater treatment itself and transportation, are associated with substances emissions into water, air or soil whereby the LCA is quantitatively processed. In the present study, the direct domestic carbon dioxide emission from the field tidal flow CW was excluded from the analysis as it is biogenic in nature while the methane and nitrous oxide involved in the calculation were gained from the previous study [15]. The extra emissions from the DAS were listed and calculated separately in order to clearly interpret the exact environmental impacts of the DAS. The existence of vegetation could be a carbon sink except for the fantastically aesthetic value. The CO_2 uptake capacity of the vegetation is between 617 and 977 g C/(m^2·year) [16] and 800 g C/(m^2·year) was adopted in the present study.

2.4. Life Cycle Impact Assessment

The LCA was performed in E-balance software [13] in accordance with the international standards ISO 14040/14044 [8]. CML2002 LCIA methodology was used in the analysis at the Midpoint level. The climate change-associated impacts and environmental quality-associated issues were the main focuses of the present LCA. Impact categories involved in this methodology include fossil depletion potential (FDP), acidification potential (AP), eutrophication potential (EP) and global warming potential (GWP). In addition, the emission of CO_2, NO_x, and SO_2 was calculated as well in the model. Regarding the environmental impacts from the vegetation, the impact value from the LCA calculation was negative while others were positive. Normalization method (normalization reference of China) was used to convert the characterization results into dimensionless scores to make them comparably [17].

2.5. Sensitivity Analysis

To evaluate to what degree the processes parameters may influence the LCA outcome of the tidal flow CW, a sensitivity analysis was conducted regarding several key materials and energy input. Herein, the electricity consumption, quantity of DAS, lifetime of containers and pumps, as well as the coverage of vegetation were considered in the sensitivity analysis. According to Garfí et al. [18] a variation of ±10% was considered for all parameters and the sensitivity coefficient was calculated using Equation (1).

$$\text{Sensitivity coefficient} = \frac{\left(\text{Output}_{high} - \text{Output}_{low}\right)/\text{Output}_{default}}{\left(\text{Input}_{high} - \text{Input}_{low}\right)/\text{Input}_{default}} \tag{1}$$

where, Input is the value of the input variable (i.e., DAS, effluent quality, chemicals and vegetation), while Output is the value of the environmental indicator (i.e., FDP, AP etc.).

3. Results and Discussion

3.1. Interpretation of the LCA Results

The characterization and normalization results of LCA regarding the field tidal CW are summarized and presented in Tables 3 and 4, respectively. Characterization results enable the implementers to figure out the most positive contributing process to the environmental degradation within each category while the normalization results present the predominant damage category within each process [17].

From the characterization results, the highest contributor to each environmental impact category is clearly shown. Pumps manufacturing, electricity generation, effluent, and tidal flow CW treatment processes dominated the AP, FDP, EP, and GWP, respectively. Moreover, the electricity generation mainly induced the CO_2 and SO_2 emissions while the DAS transportation was the main source of NO_x emission. Particularly, the electricity generation dominated three environmental impact categories and is deemed to be the key process contributing to the destruction of the environment. This is unexpectedly out of the conventional scope to CW as CW is recognized as an energy-effective technology for wastewater treatment. This is probably because of the application of tidal flow operational strategy in the field CW project to enhance the treatment efficiency while tidal CW relies on the pumps to generate the tides.

On the other hand, the normalization results also presented some important information to guide the application of the tidal flow CW in treating swine wastewater. In addition, it can help to determine the most significantly environmental impact in each process. Obviously, the most significant impact is FDP for electricity generation, container manufacturing, and pump manufacturing, NO_x emission for gravel transportation, and EP for DAS and CW performance, respectively.

Table 3. The characterization results of the present LCA.

Process	AP (kg SO_2 eq)	FDP (kg Coal-R eq)	CO_2 (kg)	EP (kg PO_4^{3-} eq)	GWP (kg CO_2 eq)	NO_x (kg)	SO_2 (kg)
Electricity	1.08×10^{-2}	5.93	1.99	5.65×10^{-4}	2.05	3.73×10^{-3}	7.92×10^{-3}
Gravel	1.33×10^{-4}	4.80×10^{-2}	1.07×10^{-2}	1.80×10^{-5}	1.14×10^{-2}	1.33×10^{-4}	3.90×10^{-5}
Containers	5.24×10^{-4}	1.18	1.26×10^{-1}	4.77×10^{-5}	1.47×10^{-1}	2.71×10^{-4}	3.16×10^{-4}
Pumps	8.50×10^{-4}	5.61×10^{-1}	1.60×10^{-1}	2.92×10^{-3}	1.74×10^{-1}	5.66×10^{-4}	4.76×10^{-4}
DAS	2.24×10^{-4}	6.35×10^{-2}	9.95×10^{-3}	1.01×10^{-4}	1.10×10^{-2}	1.10×10^{-2}	1.23×10^{-5}
CW	0	0	0	0.006715	0.22495	0	0
Vegetation	0	0	-0.0363	0	-0.0363	0	0

Table 4. The normalization results of the LCA.

Process	AP-CN-2010	FDP (Fossil Fuel)-CN-2010	CO_2-CN-2010	EP-CN-2010	GWP-CN-2010	NO_x-CN-2010	SO_2-CN-2010
Electricity	2.97×10^{-13}	3.83×10^{-13}	2.40×10^{-13}	1.50×10^{-13}	1.95×10^{-13}	1.79×10^{-13}	3.63×10^{-13}
Gravel	3.66×10^{-15}	3.10×10^{-15}	1.29×10^{-15}	4.78×10^{-15}	1.09×10^{-15}	6.41×10^{-15}	1.78×10^{-15}
Containers	1.44×10^{-14}	7.60×10^{-14}	1.52×10^{-14}	1.27×10^{-14}	1.40×10^{-14}	1.30×10^{-14}	1.45×10^{-14}
Pumps	2.33×10^{-14}	3.63×10^{-14}	1.93×10^{-14}	7.77×10^{-13}	1.65×10^{-14}	2.72×10^{-14}	2.18×10^{-14}
DAS	6.16×10^{-15}	4.11×10^{-15}	1.20×10^{-15}	2.68×10^{-14}	1.04×10^{-15}	1.45×10^{-14}	5.64×10^{-16}
CW	0	0	0	1.79×10^{-12}	2.13×10^{-14}	0	0
Vegetation	0	0	-4.37×10^{-15}	0	-3.45×10^{-15}	0	0

3.2. AP, SO_2, and NO_x Emission

AP was mainly due to the SO_2 and NO_x emissions from the fossil fuel combustion which generated electricity or fueled the transportation [19]. As aforementioned, fossil fuel is still one of the first energy sources in Ireland. Thus, electricity generation shared almost 90% contribution to SO_2 emission (Figure 2). On the other hand, pump manufacturing and DAS transportation had the biggest contribution to the AP and NO_x emission.

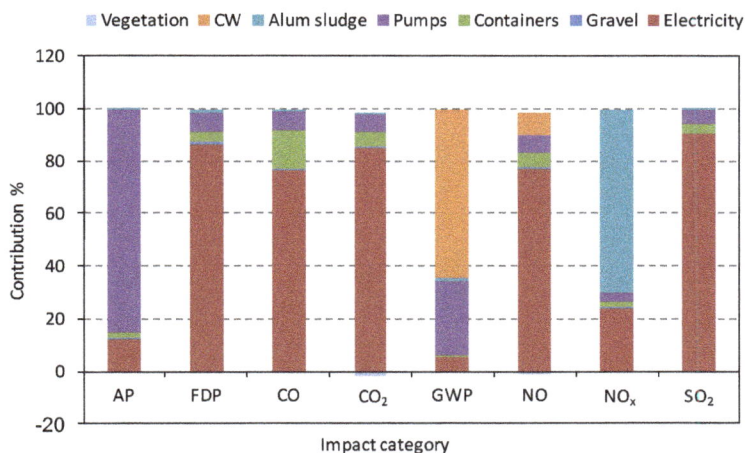

Figure 2. Contribution of all the processes to each environmental impact category.

3.3. EP

EP refers to the environmental impact caused by the nutrients released into the surface water [9]. For the wastewater treatment facilities, N and P residuals in the effluent are the main culprits of the EP. As such, the effluent from the tidal flow CW was the predominant source of EP, sharing about 64% of the impact (Figure 2). As shown in Table 1, the average N and P contents in the effluent were 71 and 1.5 mg/L, respectively. The high N residual is really a nuisance to the environment, particularly the surface water bodies. Under the circumstance of spreading the effluent on grassland, the adverse impact on groundwater should not be ignored either. According to the pollutant removal performance, the ammonia and nitrate content in the effluent were still high, which needs further improvement.

3.4. CO₂ Emission and GWP

GW is an indicator to quantify the greenhouse gas (GHG) emission of the target product. Apart from the CO_2, GWP also includes methane and nitrous oxide. Indeed, the concept of "carbon neutrality" in wastewater treatment describes the GWP. In general, conventional CW is always a carbon sink rather than a GHG source [18]. That is why nature-based solutions are always the priority for wastewater treatment.

However, the tidal flow CW in the present study showed a positive GWP and CO_2 emission which means that the tidal flow CW was not a carbon sink any more. According to Figure 2, the electricity generation shared most of the GHG emission with a proportion of 77–85%. It is worth noting that the CW also contributed about 8% to the GWP. Although the CO_2 emission from the CW was not considered, the tidal flow CW also generated a considerable quantity of the methane and nitrous oxide. In addition, CW is also acknowledged by the existence of the vegetation which could uptake the CO_2 from the air. However, in the present study, the CO_2 fixation by CW seemed fairly minor compared with the CO_2 generation from the electricity.

3.5. Sensitivity Analysis

According to the results of the sensitivity analysis (Table 5), the processes holding significant influence on the environmental impact categories include electricity, HDPE, and pump. In contrast, the environmental impact categories seemed immune to other processes. In the present analysis, the DAS was regarded as a kind of byproduct of the water treatment plants. Herein, no resource or energy was input into the production of the DAS and thus, there was insignificant sensitivity coming from the

DAS. In terms of the gravel, it is a natural material with little energy input and therefore, its variation induced marginal response from all the environmental impact categories. In the present study, the vegetation is also unlikely to influence the environmental impacts due to its insignificant mitigation compared with other processes, such as the electricity.

Table 5. The sensitivity analysis of the selected variables to the environmental impact categories.

Variable (±10%)	Sensitivity %						
	AP	FDP	CO_2	EP	GWP	NO_x	SO_2
DAS	0.018	0.008	0.004	0.009	0.004	0.060	0.001
Electricity	**0.862**	**0.762**	**0.853**	0.054	**0.772**	**0.745**	**0.903**
Vegetation	0	0	0.016	0	0.014	0	0
HDPE	0.042	**0.151**	0.054	0.004	0.055	0.054	0.036
Pump	0.068	0.072	0.068	**0.28**	0.065	**0.113**	0.054
Gravel	0.010	0.006	0.004	0.001	0.004	0.027	0.004

Note: Bold letter means the significantly sensitive category.

By the contrast consideration, all the environmental impact categories except for the EP are sensitive to the variation of the electricity. A 7–9% increase was recorded for each category with a 10% increase of the electricity consumption. This result also demonstrated again that the electricity is the dominant source of almost all the adversely environmental impacts. In addition, the variation of the HDPE also significantly influences the FDP while the variation of pump influenced the EP and NOx emission. On the other hand, the results in Table 5 indicate the remaining aspects that need further improvement to mitigate the environmental impacts.

4. Summary and Conclusions

In the present LCA analysis, a field tidal flow CW for treating a farmland wastewater was assessed based on the environmental impacts. The LCA results indicated that the present field tidal flow CW unexpectedly presented a high adverse impact on the environment. In several previous studies regarding the environmental performance of CWs [18], the CO_2 emission is at least a negative value which means that the CW sequesters carbon from the atmosphere and mitigates the adverse impacts from other materials/energy input. However, in the present field tidal flow CW, the CO_2 emission was positive due to the electricity consumption. Consequently, the CO_2 sequestration by the vegetation in the field tidal flow CW was very small compared with that emitted from the electricity generation.

On the other hand, the operation of the pumps mainly contributed to the electricity consumption in the field tidal CW. Indeed, the tidal flow CW is designed to enhance the oxygen diffusion efficiency and nitrification process. This strategy is an effective alternative to the aeration process and had led to high treatment efficiency [10]. However, from this LCA analysis, the energy consumption of the field tidal flow CW was very high, which induced the adverse environmental impacts. Thus, further comparison and discussion are needed to verify the environmental impact of tidal flow scheme and artificial aeration. It is noted that background information, e.g., the wastewater treated and the background conditions considered in the LCA, should be taken into account to fairly evaluate the environmental performance of tidal flow CW in the present LCA. The livestock farms are usually located far away from each other. Thus, it is difficult to converge all the wastewater and then treat it with conventional activated sludge systems. Generally, the cost of activated sludge systems scales down with the increase of wastewater loading rate. Therefore, it is cost-intensive to adopt the conventional activated sludge system. Regardless, the real environmental performance of activated sludge in comparison to the tidal flow CW in treatment of livestock wastewater needs further evaluation.

In addition, the footprint is another major concern in choosing the appropriate technology. The tidal flow CW is usually small compared with conventional CWs. This favorable feature of tidal flow CW extends its application in some land-limited area. However, a lot of land is available on most livestock farms. Thus, the suitability of tidal flow CW in treating livestock wastewater is rooted in its recognized treatment performance.

Overall, in this study, it has been demonstrated via a field tidal flow CW for purifying swine wastewater that the LCA seems useful for providing entire environmental impact. Although tidal flow is a novel strategy in CW technology, the sustainability of the tidal flow CW was frustrated by the high electricity consumption to create the tides. The electricity generation accounted about 60–80% of the adverse impacts in the FDP, GWP, CO_2 and SO_2 emissions. Attention should be paid to the application of the tidal flow strategy in some cases.

Author Contributions: Tong Wang and Yaqian Zhao conceived and designed the experiments; Kate O'Meara and Emmet Mullan performed the experiments; Ranbin Liu and Tong Wang analyzed the data; Tong Wang, Ranbin Liu and Yaqian Zhao wrote the paper.

Funding: This research was partially funded by National Natural Science Foundation of China [No. 41572235].

Acknowledgments: This study was partially supported by National Natural Science Foundation of China (No. 41572235).

Conflicts of Interest: The authors declare no conflict of interest.

References

1. Agriculture and Food Development Authority Agriculture in Ireland. Available online: https://www.teagasc.ie/rural-economy/rural-economy/agri-food-business/agriculture-in-ireland/ (accessed on 2 April 2018).
2. Journal Engineers Green farming: Making the Pig Industry a Nexus of Waste Management and Renewable Energy. Available online: http://www.engineersjournal.ie/2016/07/12/green-farm-pig-renewable-energy/ (accessed on 2 April 2018).
3. Portal, E.D. Nitrate Vulnerable Zones. Available online: https://www.europeandataportal.eu/data/en/dataset/nitrate-vulnerable-zones (accessed on 2 April 2018).
4. Kadlec, R.H.; Wallace, S.D. *Treatment Wetlands*, 2nd ed.; Taylor & Francis Group: Boca Raton, FL, USA, 2009; ISBN 978-1-56570-526-4.
5. Hu, Y.; Zhao, Y.; Rymszewicz, A. Robust biological nitrogen removal by creating multiple tides in a single bed tidal flow constructed wetland. *Sci. Total Environ.* **2014**, *470–471*, 1197–1204. [CrossRef] [PubMed]
6. Hu, Y.; Zhao, X.; Zhao, Y. Achieving high-rate autotrophic nitrogen removal via Canon process in a modified single bed tidal flow constructed wetland. *Chem. Eng. J.* **2014**, *237*, 329–335. [CrossRef]
7. Zang, Y.; Li, Y.; Wang, C.; Zhang, W.; Xiong, W. Towards more accurate life cycle assessment of biological wastewater treatment plants: A review. *J. Clean. Prod.* **2015**, *107*, 676–692. [CrossRef]
8. Corominas, L.; Foley, J.; Guest, J.S.; Hospido, A.; Larsen, H.F.; Morera, S.; Shaw, A. Life cycle assessment applied to wastewater treatment: State of the art. *Water Res.* **2013**, *47*, 5480–5492. [CrossRef] [PubMed]
9. Kalbar, P.P.; Karmakar, S.; Asolekar, S.R. Assessment of wastewater treatment technologies: Life cycle approach. *Water Environ. J.* **2013**, *27*, 261–268. [CrossRef]
10. Zhao, Y.Q.; Babatunde, A.O.; Hu, Y.S.; Kumar, J.L.G.; Zhao, X.H. Pilot field-scale demonstration of a novel alum sludge-based constructed wetland system for enhanced wastewater treatment. *Process Biochem.* **2011**, *46*, 278–283. [CrossRef]
11. Remy, C.; Boulestreau, M.; Warneke, J.; Jossa, P.; Kabbe, C.; Lesjean, B. Evaluating new processes and concepts for energy and resource recovery from municipal wastewater with life cycle assessment. *Water Sci. Technol.* **2015**, *73*, 1074–1080. [CrossRef] [PubMed]
12. Yang, Y.; Zhao, Y.Q.; Babatunde, A.O.; Wang, L.; Ren, Y.X.; Han, Y. Characteristics and mechanisms of phosphate adsorption on dewatered alum sludge. *Sep. Purif. Technol.* **2006**, *51*, 193–200. [CrossRef]
13. Bai, S.; Wang, X.; Huppes, G.; Zhao, X.; Ren, N. Using site-specific life cycle assessment methodology to evaluate Chinese wastewater treatment scenarios: A comparative study of site-generic and site-specific methods. *J. Clean. Prod.* **2017**, *144*, 1–7. [CrossRef]
14. Sustainable Energy Authority of Ireland. Energy in Ireland 1990–2015. Available online: http://www.seai.ie/resources/publications/Energy-in-Ireland-1990-2015.pdf (accessed on 3 April 2018).
15. Mander, Ü.; Maddison, M.; Soosaar, K.; Karabelnik, K. The impact of pulsing hydrology and fluctuating water table on greenhouse gas emissions from constructed wetlands. *Wetlands* **2011**, *31*, 1023–1032. [CrossRef]

16. De Klein, J.J.M.; Van der Werf, A.K. Balancing carbon sequestration and GHG emissions in a constructed wetland. *Ecol. Eng.* **2014**, *66*, 36–42. [CrossRef]
17. Bai, S.; Wang, X.; Zhang, X.; Zhao, X.; Ren, N. Life cycle assessment in wastewater treatment: Influence of site-oriented normalization factors, life cycle impact assessment methods, and weighting methods. *RSC Adv.* **2017**, *7*, 26335–26341. [CrossRef]
18. Garfí, M.; Flores, L.; Ferrer, I. Life Cycle Assessment of wastewater treatment systems for small communities: Activated sludge, constructed wetlands and high rate algal ponds. *J. Clean. Prod.* **2017**, *161*, 211–219. [CrossRef]
19. Corbala-robles, L. Life cycle assessment of biological pig manure treatment versus direct land application—A trade-off story. *Resour. Conserv. Recycl.* **2018**, *131*, 86–98. [CrossRef]

water MDPI

Article

Development of an Integrated Modelling System for Evaluating Water Quantity and Quality Effects of Individual Wetlands in an Agricultural Watershed

Yongbo Liu [1], Wanhong Yang [1,*], Hui Shao [1], Zhiqiang Yu [2] and John Lindsay [1]

[1] Department of Geography, University of Guelph, 50 Stone Road E., Guelph, ON N1G 2W1, Canada; lyongbo@uoguelph.ca (Y.L.); shaoh@uoguelph.ca (H.S.); jlindsay@uoguelph.ca (J.L.)
[2] Civica Infrastructure, Vaughan, ON L6A 4P5, Canada; hawklorry@gmail.com
* Correspondence: wayang@uoguelph.ca; Tel.: +1-519-824-4120 (ext. 53090)

Received: 3 May 2018; Accepted: 11 June 2018; Published: 13 June 2018

Abstract: A GIS-based fully-distributed model, IMWEBs-Wetland (Integrated Modelling for Watershed Evaluation of BMPs—Wetland), is developed to simulate hydrologic processes of site-specific wetlands in an agricultural watershed. This model, powered by the open-source GIS Whitebox Geospatial Analysis Tools (GAT) and advanced database technologies, allows users to simulate and assess water quantity and quality effects of individual wetlands at site and watershed scales. A case study of the modelling system is conducted in a subbasin of the Broughton's Creek Watershed in southern Manitoba of Canada. Modelling results show that the model is capable of simulating wetland processes in a complex watershed with various land management practices. The IMWEBs-Wetland model is unique in simulating the water quantity and quality effects of individual wetlands, which can be used to examine location-specific targeting of wetland retention and restoration at a watershed scale.

Keywords: distributed watershed modelling; individual wetlands; wetland retention and restoration; water quantity and quality; location-specific targeting; agricultural watersheds

---

1. Introduction

Wetland retention and/or restoration is an important best management practice (BMP) in agricultural watersheds as it provides critical hydrologic functions including flood attenuation, groundwater recharge, and contaminant filtering. Various hydrologic models have been applied to study wetland processes and evaluate their impacts on water quantity and quality at a watershed scale. However, most of these models have a semi-distributed structure, which lump all wetlands in a subbasin into a single functionally equivalent wetland for parameterization [1]. Due to the complex and dynamic inter-connections of wetlands within a watershed, it is challenging to characterize the spatial heterogeneity of wetlands with a semi-distributed model. In recent years, several efforts have been made to address this challenge in watershed hydrologic modelling. These model enhancements have contributed to an improved understanding of wetland functions and services at a watershed scale.

Wang et al. [2] proposed a hydrologic equivalent wetland (HEW) concept in the Soil and Water Assessment Tool (SWAT) to characterize multiple wetlands within a subbasin. This approach lumps multiple wetlands within a subbasin into one wetland based on the aggregation of their geometric characteristics including wetland surface area, volume, and drainage area. The HEW function is achieved through calibrating wetland parameters including the fraction of the subbasin area that drains into wetlands, the volume of water stored in the wetlands when filled to their normal water level, the volume of water stored in the wetlands when filled to their maximum water level, the longest tributary channel length in the subbasin, and the Manning's n value for tributary and main channels. The HEW approach has the advantage of characterizing the non-linear functional relations between runoff and wetlands [3] and is reasonable for simulating hydrological processes of prairie pothole wetlands [4,5]. Different from the HEW approach, Mekonnen et al. [6] proposed a probability distribution routine in the SWAT model to represent multiple wetlands within a subbasin. This approach applied a probability density function to characterize the spatial heterogeneity of landscape depression storages in a watershed and incorporated the seasonal variation of soil erodibility to account for the change of erosion rate during soil freeze and thaw. An application of the approach showed an improved simulation of sediment export in a Canadian prairie watershed [7].

Nasab et al. [8] developed an alternative depression characterization approach in the SWAT model to characterize multiple wetlands within a subbasin. This approach utilizes topographic characteristics and distribution of depressions to establish hierarchical relationships of depressions. This approach generates detailed hydro-topographic characteristics of depressions to parameterize SWAT pothole features by merging lower level depressions into higher level depressions. An application of the approach to a watershed with numerous potholes in North Dakota showed an improved model performance in simulating stream flows at the watershed outlet. Another approach in enhancing SWAT characterization of wetlands is to couple the SWAT model with a United States Environmental Protection Agency (USEPA) field scale model called System for Urban Stormwater Treatment and Analysis Integration (SUSTAIN) that characterizes drained water from upland flow [9,10]. In addition to wetland modules developed in SWAT, improvements have also been made in the semi-distributed Soil and Water Integrated Model (SWIM) [11,12] and the distributed hydrological model HYDROTEL [13,14] through incorporating wetland flow, nutrient, and groundwater related dynamics to characterize wetland processes.

While the semi-distributed watershed models have had various enhancements and applications, there still exists the limitation on spatially explicit characterization of wetlands and their connectivity. In the SWAT model, areas with the same land use, soil and slope class in a subbasin are grouped into hydrologic response units (HRUs) to characterize their spatial heterogeneity. However, HRUs are not spatially connected within a subbasin. Evenson et al. [15,16] developed an alternative approach to characterize geographically isolated wetlands (GIWs) by redefining SWAT HRUs to conform with the mapped GIWs and their drainage boundaries. New model input files were constructed to direct the simulation of GIW fill-spill hydrology and upland flows to GIWs. The enhanced SWAT was applied to a North Carolina watershed and a North Dakota watershed to examine wetland effects on stream flow, baseflow and peak flows with a satisfactory model performance at the watershed outlet. Modelling application to a coastal plain watershed in North Carolina showed that increased extent of isolated and riparian wetlands had significant effects on decreasing seasonal and annual flows [17]. This approach of wetland representation in the SWAT model contributes to a more realistic simulation of wetland effects. However, the lumping at the subbasin level due to inherent semi-distributed model structure prevents the examination of wetland effects at specific locations. There is a need to develop a fully distributed watershed model to explicitly characterize individual wetlands and simulate their effects on flow and water quality at a watershed scale.

In the period of 2004–2013, Agriculture and Agri-Food Canada's (AAFC) Watershed Evaluation of BMPs (WEBs) program conducted BMP assessments in nine experimental watersheds across Canada [18]. In the WEBs program, the Guelph Watershed Evaluation Group developed a cell-based integrated modelling system for watershed evaluation of BMPs (IMWEBs) to conduct BMP assessments at site, field, farm, and watershed scales [19]. The foundation of IMWEBs was the Water and Energy Transfer between Soil, Plant and Atmosphere (WetSpa) model, a fully distributed hydrologic model for flood prediction and watershed management [20]. Several other well-known agricultural watershed models such as SWAT [21], Agricultural Policy Environmental Extender (APEX) [22] and Riparian Ecosystem Management Model (REMM) [23] were also referenced in the IMWEBs development. The IMWEBs model has an object-oriented and modular-based structure to perform dynamic watershed modelling supported by five databases including geospatial, hydro-climate, BMP, parameter, and model output. Hydrologic processes simulated in the model include climate, snowmelt, water balance, plant growth, soil erosion, nutrient cycle, and channel routing, while general agricultural BMPs including crop management, fertilizer management, and tillage management are incorporated in the model simulation. The IMWEBs model was firstly applied to the 75 km² South Tobacco Creek watershed in southern Manitoba of Canada and satisfactory modelling results were obtained [19].

The objective of this paper is to present the development of a wetland module as one of the BMP components in the IMWEBs model (IMWEBs-Wetland). A case study in a small agricultural watershed is also presented to demonstrate its applicability, performance, and usefulness in characterizing wetlands in an agricultural watershed. With a cell-based structure, the IMWEBs-Wetland model is specifically designed for simulating and evaluating water quantity and water quality effects of individual wetlands at a watershed scale. This makes the model more convenient and straightforward for supporting site-specific wetland retention and restoration. The modelling system also has novelties of using open-source GIS and databases and a modular structure to characterize different hydrologic processes. An object-oriented computer interface is developed to facilitate watershed delineation, model setup, model parameterization, scenario design, and output analysis. The model was applied to a 15.7-km² watershed in southern Manitoba of Canada to evaluate wetland effects. In addition to simulating the hydrologic processes operating in each wetland, the IMWEBs-Wetland model can also produce spatial distributions of various hydrologic variables at user-defined spatial and temporal scales. This makes the model an effective tool for spatial watershed management, particularly the assessment of site-specific wetland retention and restoration scenarios. Discussions of the model performance and future development perspectives are provided at the end of this paper.

2. Model Development

2.1. System Design

In addition to other existing modules in the IMWEBs modular library, IMWEBs-Wetland is designed to simulate and assess wetland hydrologic processes at both individual and watershed scales. Processes simulated in the wetland module include water balance, sediment balance, and nutrient balance. In this study, wetlands that are isolated from mainstreams and dominated by fill and spill processes [1] are simulated in the IMWEBs-Wetland model, while riparian wetlands along mainstreams are simulated with channel processes in the model.

In the IMWEBs-Wetland model, a watershed is delineated into subbasins, and one subbasin contains one isolated wetland at the subbasin outlet. These isolated wetlands are spatially connected in the model through a DEM delineated flow path once they are filled. Other subbasin outlets can be defined at major tributary confluences, monitoring stations, watershed outlet, and user defined locations assisted by the modelling interface. In addition to the general IMWEBs inputs including climate, DEM, soil, land use, and land management, wetland inventory data with associated wetland parameters are required for model setup. Wetland parameters such as surface area, volume, and drainage area are estimated based on available DEM and wetland inventory data with the IMWEBs-Wetland interface prior to model simulation. Outputs from the IMWEBs-Wetland model include time series of flow, sediment, and nutrient concentration at any user defined wetland or reach locations, and spatial distribution of wetland or watershed hydrologic variables at user-defined spatial and temporal scales determined during model setup. A diagram for a delineated wetland subbasin characterized in the IMWEBs-Wetland model is shown in Figure 1.

To follow the modelling structure of the IMWEBs model, cells within a wetland polygon are simulated the same as upland cells and are not grouped into one unit. Processes of runoff and pollutant generation for cells within the wetland polygon are simulated using IMWEBs algorithms based on the land cover and DEM data. However, soil parameters, particularly the wetland bottom hydraulic conductivity are modified during IMWEBs-Wetland model setup. As such, the wetland module is on the top of the existing IMWEBs model with inputs from its drainage areas, and outputs the same as those of the reservoirs. Wetlands are connected through surface water and groundwater in the model. Surface water flows along the pathway derived from the DEM once the wetland is filled or above its normal storage for drained–altered wetlands. Groundwater is simulated separately from the wetland module at a subbasin scale using a non-linear reservoir method based on its contributing area. Descriptions on groundwater simulation in the model are provided in Sections 2.3 and 2.5.

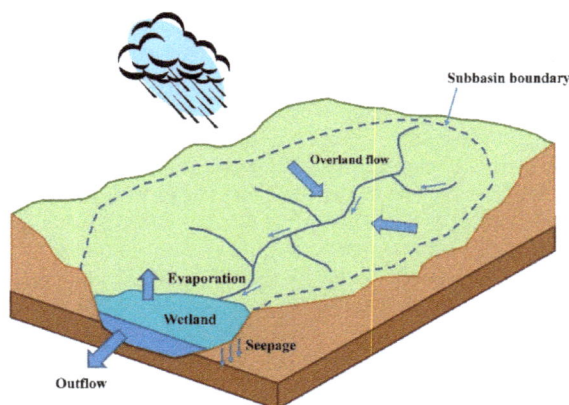

Figure 1. An open-water wetland and its drainage area delineated in the IMWEBs-Wetland model.

2.2. Wetland Classification and Characterization

Five types of wetlands including drained–altered, drained–consolidated, drained–lost, undrained–altered, and undrained–intact are classified and simulated in the IMWEBs-Wetland model based on the Ducks Unlimited Canada (DUC) wetland inventory data (Figure 2). The drained–altered wetland has an outlet drain, while drained–consolidated wetland has an inlet drain, both with riparian and aquatic vegetation present. The drained–lost wetland has an outlet drain without riparian and aquatic vegetation present. The undrained–altered wetland does not have drains with riparian and aquatic vegetation absent or disturbed, while undrained–intact wetland does not have drains with riparian and aquatic vegetation present. The drained–lost and drained–altered wetlands are similar in that they are all drained but one is altered for cultivation and the other is vegetated with native plants. This multi-temporal classification approach defines wetlands that are in either drained or undrained state. The wetland classification provides a basis for identifying existing wetlands for retention and drained–lost wetlands for restoration during IMWEBs-Wetland scenario assessment.

The drained–consolidated and undrained–intact wetlands have similar geometric features for which no outlet drains exist, and flow out of the wetlands follows the fill-and-spill process. For these two types of wetlands, the normal wetland surface area and normal storage are assumed equal to the maximum wetland surface area and maximum wetland storage. The following volume–area regression equations are used to estimate wetland storage based on the wetland surface area [24]:

$$V = 2.85A^{1.22} \text{ when A } \leq \text{ 70 ha} \tag{1}$$

$$V = 7.1A + 9.97 \text{ when A } > \text{ ha,} \tag{2}$$

where A is the wetland surface area (ha) and V is the corresponding full supply volume (10^3 m^3). The constant in Equations (1) and (2) can be adjusted for specific wetlands if observation data are available. Discharge out of the wetland is calculated after the wetland is filled, and all excess water discharges into the downstream reach during the time step.

The drained–altered wetland is different from above in that the wetland is drained and altered but still maintain water in the wetland with a reduced holding capacity. Flow out of the wetland occurs when water level in the wetland is higher than outlet drains. To characterize this type of wetland, the maximum storage is calculated using Equations (1) and (2) based on the wetland inventory data, while the normal wetland surface area is assumed to be 1/3 of the maximum surface area if no field survey data are available, and the corresponding normal storage is calculated using Equations (1) and (2). Discharge out of the wetland when water volume is between normal storage and maximum storage is calculated using equation (3) from the SWAT model [25]:

$$V_{flowout} = (V_1 + V_{flowin} - V_{normal})/C,\tag{3}$$

where V_{flowin} and $V_{flowout}$ are the water volume entering and flowing out of the wetland over the time step (m^3), V_1 is the end storage of previous time step (m^3), V_{normal} is the wetland normal storage (m^3), and C is a wetland discharge coefficient with a default value of 10. No outflow occurs if sum of inflow and existing storage is less than normal wetland storage. When calculated wetland water volume is over maximum storage, all excess water discharges during the time step. The wetland discharge coefficient can be adjusted for specific wetlands when field observation data are available.

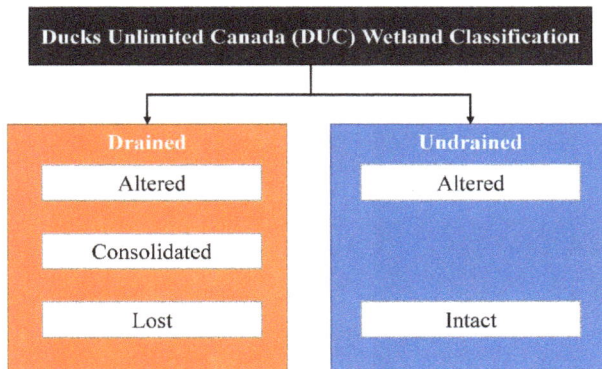

Figure 2. DUC wetland classification.

2.3. Water Balance

The water balance for a wetland is:

$$V_2 = V_1 + V_{flowin} - V_{flowout} + \Sigma(V_{pcp} - V_{evap} - V_{seep})_{wetland\_cells},\tag{4}$$

where V_2 is the water volume in the wetland at the end of the time step (m^3), V_{pcp} is the precipitation volume on the wetland over the time step (m^3), V_{evap} is the evaporation/evapotranspiration from the wetland surface over the time step (m^3), and V_{seep} is the water volume lost from the wetland by seepage (m^3). To avoid duplicate calculation, V_{pcp}, V_{evap}, and V_{seep} are calculated in the IMWEBs-Wetland model and summed for cells within the wetland polygon. For each wetland cell, based on land cover and wetland inventory data, V_{evap} is estimated with a potential evaporation/evapotranspiration rate in which evaporation is calculated for wetlands with an open water surface, and evapotranspiration is calculated for wetlands with a vegetation cover including forest. V_{seep} is calculated based on wetland bottom conductivity and water availability calculated in the wetland module.

Drained–consolidated wetland is different from drained–altered or drained–lost wetlands in that water is drained through surface inlet drains connected to a tile underlying the wetland rather than at the wetland outlet. To simulate this process in the model, flow at the subbasin outlet is calculated by summing surface runoff from non-wetland cells within the subbasin and bypasses the wetland into the downstream reach. Water balance in the consolidated wetland is maintained with inputs of precipitation in the wetland area and lateral flow from upland fields, while outflow from the wetland occurs when the wetland is filled, and evapotranspiration and seepage are calculated the same as those for drained–altered wetlands. Output from the wetland including flow, sediment and nutrient loading replaces the original reach output of the subbasin and joins the routing of downstream reaches.

2.4. Sediment and Nutrient Balance

A mass balance approach in the SWAT model [25] is used for wetland sediment and nutrient routing. The governing equation is:

$$M_2 = M_1 + M_{flowin} - M_{flowout} - M_{stl}. \tag{5}$$

For sediment balance, M_2 and M_1 are the amount of sediment in the wetland at the end and at the beginning of the time step (ton), M_{flowin} is the amount of sediment added to the wetland with inflow (ton), $M_{flowout}$ is the amount of sediment transported out of the wetland with outflow (ton), and M_{stl} is the amount of sediment removed from the water body by settling (ton). M_{flowin} is obtained from IMWEBs-Wetland reach output of the subbasin. As the wetland is located at the outlet of a subbasin, all cells within the subbasin including wetland area contribute runoff, sediment and nutrient yields at the subbasin outlet. Calculation of sediment deposition is based on actual sediment concentration and equilibrium sediment concentration of the wetland using the SWAT approach [25]. $Sed_{flowout}$ is the suspended sediment out of the wetland calculated by outflow multiplied by sediment concentration. The calculated sediment loading replaces the reach sediment output and joins the routing of downstream reaches. This method assumes a uniform distribution of deposited sediment on the wetland bottom without accounting for its spatial distributions.

The SWAT wetland water quality algorithms are used to simulate nutrient balance for each wetland. The mass balance for nitrogen and phosphorous is similar as the sediment mass balance, where M_2 and M_1 in Equation (5) are the amount of nutrients in the wetland at the end and at the beginning of the time step (kg), M_{flowin} is the amount of nutrients added to the wetland with inflow (kg), $M_{flowout}$ is the amount of nutrients transported out of the wetland with outflow (kg), and M_{stl} is the amount of nutrients removed from the water body by settling and seepage (kg). Detailed descriptions of the method can be found in [25].

2.5. Groundwater

In the IMWEBs model, baseflow is calculated at subbasin scale using a non-linear reservoir method. This approach is applicable for a relatively large subbasin. However, when delineated subbasin is very small, the groundwater flow may not contribute to the subbasin outlet but at a location in downstream reaches. This pattern causes challenge for groundwater characterization in the IMWEBs-Wetland model, because individual wetlands are located at subbasin outlets, and their drainage areas may be very small for a prairie watershed with numerous pothole wetlands. As a result, groundwater may bypass the outlet from underground and does not join the flow at the subbasin outlet.

To solve this problem, a threshold drainage area (ha) is incorporated in the IMWEBs-Wetland model for groundwater simulation at the reach outlet. This threshold area can be estimated based on field observations at a site where groundwater flow is initiated during snowmelt period or after heavy storms. It can be also determined through model calibration if flow data are available at monitoring stations with small contribution areas within the watershed. If the reach contribution area is less than the threshold value, no groundwater outflow exists at the reach outlet, and the calculated groundwater flow for this subbasin is accumulated to the next downstream reach. Once the reach drainage area is greater than the threshold value, the accumulated groundwater flow returns to the channel. The groundwater flow calculated for this subbasin and the groundwater flow accumulated from upstream subbasins are summed and added to the inflow of the river reach for channel routing. These assumptions are workable for upstream subbasins based on reach drainage areas calculated during model setup. For small subbasins in middle and downstream areas adjacent to the main channel, if the subbasin area is less than the threshold value, the calculation follows the same way as upstream subbasins, and the calculated groundwater flow is added to the main channel outlet.

3. Interface Development

3.1. Framework and Database Management

A computer interface was developed to assist IMWEBs-Wetland input data preparation, watershed delineation, model setup, parameterization, and result visualization. The interface, powered by the Whitebox Geospatial Analysis Tools (GAT) and SQLite database technologies, facilitates the user to simulate water quantity and quality effects of individual wetlands at site, field, farm and watershed scales. Whitebox GAT is an open-source desktop GIS and remote sensing software package for general applications of geospatial analysis and data visualization and is intended to provide a platform for advanced geospatial data analysis with applications in both environmental research and the geomatics industry [26]. The purpose of the interface is to conduct pre- and post-processing for the IMWEBs-Wetland model. These include: (a) to delineate watershed subbasins accounting for each wetland; (b) to compute model parameters for each wetland; (c) to display wetland and watershed results; and (d) to manage all associated wetland and watershed input, output, and parameter databases.

The IMWEBs-Wetland model has five databases and one modular library (Figure 3). The geospatial database is a collection of geometric entities and attributes of the watershed such as DEM, soil, land use, streams, boundaries, climate and hydrologic stations, wetlands, reservoirs, and hydraulic structures. The hydro-climate database is a collection of climate, flow, and water quality observation data used for model input and model calibration. The BMP database stores and manages current and future BMP scenarios including BMP types, distribution, and associated parameters. The model parameter database is a collection of model parameters estimated from the geospatial and the BMP database or prepared during model development. The modular library is a collection of modules/algorithms to support a modelling exercise for a specific modelling objective and is the key component of the IMWEBs-Wetland model. Each module is self-contained and is designed to simulate a specific hydrologic process. One process may have several simulation modules depending upon the user's selection during model setup. The model output database stores model outputs including time series of flow, sediment, and water quality at user defined locations and spatial distribution of hydrologic variables at user defined spatial and temporal scales. Particularly, the IMWEBs-Wetland model provides time series and spatial distribution of wetland water balance, sediment balance, and nutrient balance at both site and watershed scales. Outputs of the IMWEBs-Wetland model can be exported to text files, SQLite databases, or both. Text file outputs are easier for editing and viewing but take more time and space, while SQLite database provides more flexibility for the interface to search, summarize, and analyze wetland results.

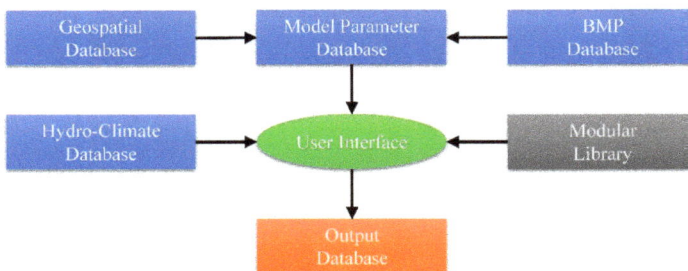

Figure 3. Structure of the IMWEBs-Wetland model.

3.2. Drainage Delineation

To simulate the hydrologic processes of individual wetlands, their contribution areas and the drainage network that links each wetland need to be properly defined to characterize interconnections of wetlands during model simulation. Traditionally, watershed drainage delineation requires a filling algorithm to remove depressions and flat areas in the watershed so that a continuous stream network can be generated using a GIS. This approach eliminates the depression information, e.g., wetland, in a grid DEM, which is not appropriate for wetland drainage delineation.

In the IMWEBs-Wetland model interface, a customized watershed delineation method was developed to solve this problem. Two types of input data, wetland boundary and wetland outlet (optional), are incorporated in the delineation process. If wetland outlet location data are not available, the interface can automatically generate the outlet information using the priority-flood operation approach [27] based on the DEM. For a large wetland with surface area greater than a user defined threshold value, multiple outlets are allowed for an individual wetland during the delineation process. The wetland boundary and outlet vector layers are firstly rasterized into grids containing wetland identifier values. Combining with the DEM, a D8 flow-direction raster is generated and is used to calculate flow accumulation and generate a drainage network.

Figure 4 illustrates how this updated flow direction raster is created based on the DEM, wetland boundary, and wetland outlet information. For non-boundary cells (white cells in Figure 4), flow directions are determined by examining grid elevation values using the D8 method [28], i.e., flow direction is pointed to the steepest downward neighbor cell based on elevation. By overlaying wetland boundary with the DEM, wetland boundary cells (light gray cells in Figure 4) are detected. For these wetland boundary cells, flow directions are enforced to circle the wetland within these boundary cells until a wetland outlet cell (dark gray cell in Figure 4) is reached. The subdivided wetlands within the original large wetland are treated as individual wetlands with their own outlets and flow pathways to the downstream reaches in the IMWEBs-Wetland model. Because the model generates spatial distribution results for each wetland, outputs of these subdivided wetlands are summarized automatically back after model simulation using the weighted average approach based on their surface areas within the original large wetland.

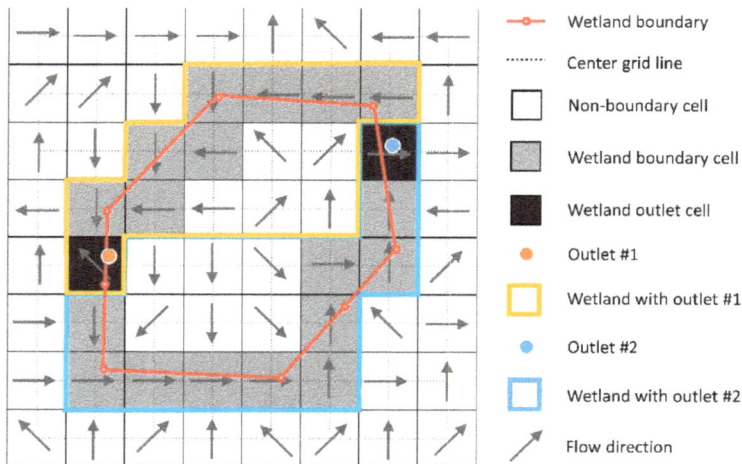

Figure 4. A schematic of wetland delineation with multiple outlets.

3.3. Wetland Parameterization

Parameters for each wetland are estimated based on wetland inventory data and are prepared after watershed delineation. Table 1 summarizes the wetland parameter name, units and their descriptions. Wetland ID, type, and maximum surface area are read from the wetland inventory attribute table. The parameter of operation year is used in the IMWEBs-Wetland model for simulating wetland loss and restoration scenarios. Parameters of subbasin ID and contribution area are obtained from the watershed delineation results. Normal water volume and maximum water volume are calculated using Equations (1) and (2), while normal water volume corresponds to the normal storage for drained–altered wetlands over which spill flow would occur. Other parameters are read from the default wetland parameter table in the BMP database. All these parameter values can be adjusted during model calibration or re-edited if field observation data are available.

Table 1. Wetland parameters for each wetland.

Parameter Name	Unit	Description
ID	-	Wetland ID
Operation	-	Year of wetland operation
Subbasin	-	Subbasin ID
Type	-	Wetland type
ContributionArea	ha	Contribution area of the wetland
NormalArea	ha	Wetland surface area at normal storage
NormalVolume	10^4 m^3	Wetland water volume at normal storage
MaxArea	ha	Wetland surface area at maximum storage
MaxVolume	10^4 m^3	Wetland water volume at maximum storage
Wet_K	mm/h	Wetland bottom saturated hydraulic conductivity
SedimentConEqui	mg/L	Wetland sediment equilibrium concentration
D50	μm	Inflow sediment median particle size
SettVolN	m/year	Nitrogen settling velocity
SettVolP	m/year	Phosphorous settling velocity
ChlaProCo	-	Chlorophyll production coefficient
WaterCalCo	m	Water clarity coefficient
InitialVolume	10^4 m^3	Initial wetland water volume
InitialSediment	mg/L	Initial wetland sediment concentration
InitialNO$_3$_mgL	mg/L	Initial wetland NO$_3$ concentration
InitialNO$_2$_mgL	mg/L	Initial wetland NO$_2$ concentration
InitialNH$_3$_mgL	mg/L	Initial wetland NH$_3$ concentration
InitialSolP_mgL	mg/L	Initial wetland soluble phosphorous concentration
InitialOrgN_mgL	mg/L	Initial wetland organic nitrogen concentration
InitialOrgP_mgL	mg/L	Initial wetland organic phosphorous concentration
RoutingConstant	-	Controlling constant for outflow routing

4. Case Study

4.1. Study Area

A case study of IMWEBs-Wetland modelling was conducted in a 15.7 km^2 small watershed, which is a subbasin of the Broughton's Creek watershed in southern Manitoba of Canada (Figure 5). The Broughton's Creek flows southeasterly into the Little Saskatchewan River, joining the Assiniboine River, and eventually entering Lake Winnipeg. The watershed has an average slope of 1.40% with a range from 0.0% to 10.8% based on a 30 m DEM and is dominated by the moderately well drained Newdale soils formed in calcareous, loamy glacial till of limestone, granite and shale origin. Agriculture is the major land use in the watershed (75.3%) with dominant crop types of spring wheat and canola. Other land use types include grassland (3.4%), wetland (20.4%), road (0.7%), and deciduous forest (0.2%). The study watershed has hundreds of undrained depressions ranging from potholes to large sloughs. Compared with potholes, sloughs are relatively large, shallow, and typically elongated northwest to southeast. In addition, there are also several small lakes present. Limited by available detailed information, this study did not differentiate potholes, sloughs, and lakes, and modeled them as wetlands. Based on the DUC wetland inventory data, a total of 492 wetlands were identified in the study watershed including 293 existing (Figure 6) and 199 drained–lost. Among the 293 existing wetlands, 85 were identified as drained–altered, 11 drained–consolidated, 133 undrained–altered, 46 undrained–intact, and 18 riparian wetlands. The total wetland area including drained–lost is 550 ha, which means 35.0% of the watershed was covered by wetland before agricultural development. Detailed descriptions of their surface areas and volumes are provided in Table 2.

The study area has a semi-arid climate, with a pronounced seasonal variation in temperature and precipitation. Based on the 1981–2010 climate data recorded at Strathclair station, located about 10 km northwest of the study watershed (Figure 5), the average yearly temperature was 1.5 °C, with the highest monthly temperature of 17.7 °C in July and the lowest monthly temperature of -17.1 °C in January. Average annual precipitation was 475 mm, of which 118 mm (25%) was snowfall, lasting from November to the following April. The average annual daily discharge at the watershed outlet was 0.065 m^3/s, ranging from 0.00 to 1.57 m^3/s based on the observed data collected at the EC9 station (Figure 5) from 2009 to 2013. The average annual runoff was 140 mm, with an average runoff coefficient of 0.29. More than 80% of this runoff and all annual peak discharges were observed in spring (late April to early May) over the monitoring period because of snowmelt. Baseflow was a small portion of the total runoff (<10%) and provided little contribution to the total flow and sediment transport.

The study watershed has suffered flooding and nutrient loading problems during spring snowmelt and heavy summer storms due to activities of wetland drainage, road construction, and land clearing for agricultural production. Therefore, retention of existing wetlands and restoration of drained–lost wetlands are important for addressing these environmental problems. The entire Broughton's Creek watershed had been selected to study the effects of wetland loss and restoration on stream water quality using the SWAT model [4,5]. However, due to the SWAT's semi-distributed model structure, the assessment of wetland loss and restoration effects was conducted at a subbasin scale but not for individual wetlands.

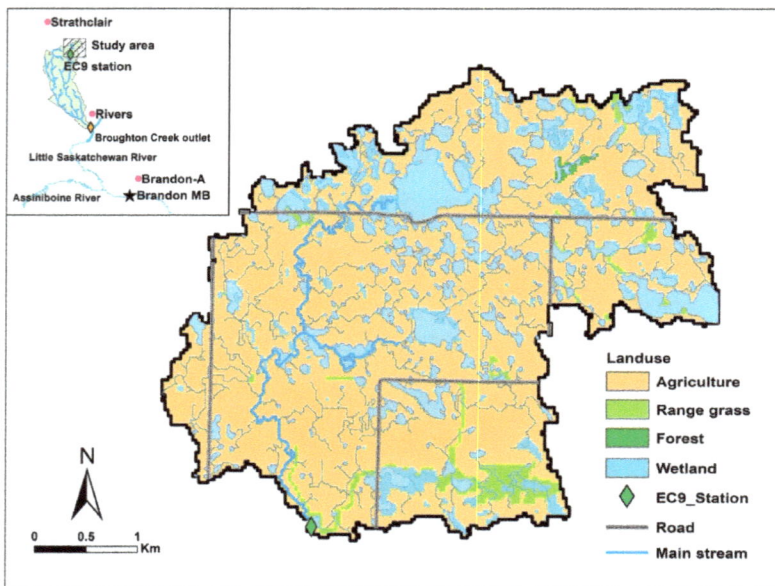

Figure 5. Location and land use of the study watershed.

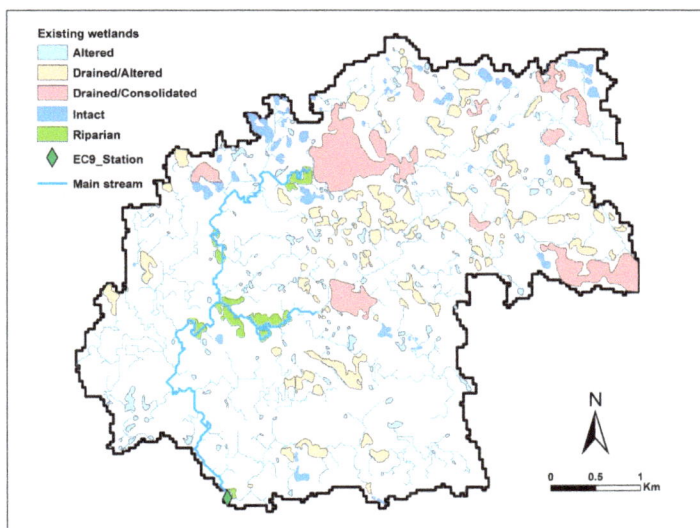

Figure 6. Existing wetlands in the study watershed.

Table 2. A summary of wetland type, number, area, and volume in the study watershed.

Wetland Type	Wetland		Surface Area		Volume	
	Number	(%)	(ha)	(%)	(10^4 m^3)	(%)
Existing, isolated, drained–altered	85	17.3	122	22.2	41.2	20.5
Existing, isolated, drained–consolidated	11	2.2	106	19.3	56.4	28.1
Non-existing, isolated, drained–lost	199	40.4	180	32.8	59.1	29.4
Existing, isolated, undrained–altered	133	27.0	53.3	9.7	13.5	6.8
Existing, isolated, undrained–intact	46	9.4	38.3	6.9	11.5	5.7
Existing, riparian	18	3.7	50.0	9.1	19.0	9.5
Total	492	100	550	100	201	100

4.2. Model Setup and Calibration

The IMWEBs-Wetland model for the study watershed was setup based on the geospatial data of DEM, land use, soil, and wetland inventory obtained from the DUC. A total of 515 subbasins were delineated, of which 511 have wetlands at the subbasin outlets and four are in the mainstreams. The original land use layer has 15 land use classes ranging from agricultural cropland to roads and trails. To convert this land use layer into IMWEBs-Wetland format, a lookup table was created linking original land use categories to the IMWEBs-Wetland land use code. Accordingly, a user-defined soil parameter database was developed based on available soil attribute data. Crop management is one of the key factors in controlling runoff, sediment and nutrient yields from an agricultural watershed. Nitrogen (N) and phosphorus (P) from agricultural land are major sources to the receiving wetlands and streams. Because no detailed crop management data were available, a two-year representative crop rotation of spring wheat and canola was assumed for the study watershed (Table 3). Crop management parameters including seeding and harvest date, fertilizer application rate and date, and tillage type and date were referenced from available literature values [29,30].

The calibration period for the study watershed was from 2009 to 2013 at a daily scale, whereas the period from 2000 to 2008 was used for warming up. This period was selected for calibration because observed flow and water quality data were available at the subbasin outlet EC9 station (Figure 5).

Precipitation and temperature data over the simulation period were obtained from the Strathclair and Rivers stations, while wind speed and wind direction data were obtained from the Brandon-A station (Figure 5). Data of solar radiation and relative moisture for the study area were downloaded from NASA's online database [31]. The solar radiation data was further validated using an image from Energy, Mines, and Resources Canada [32].

Table 3. Crop management practices in the study watershed.

Year	Crop	Practice
1	Spring wheat	Seeding on 15/5, harvest on 20/8, 78 kg/ha of N and 32 kg/ha of P on 15/5, and tillage of harrow packers on 18/5.
2	Canola	Seeding on 15/5, harvest on 20/9, 88 kg/ha of N and 32 kg/ha of P on 15/5, and tillage of light duty cultivator with harrows on 15/5 and harrow packers on 20/9.

A manual calibration was conducted for those parameters deemed most sensitive based on a parameter sensitivity analysis of the IMWEBs-Wetland model. These include runoff and water balance parameters (interception capacity for different land covers, evapotranspiration correction factor, interflow scaling factor, field capacity and porosity of top soil layer, baseflow constant and exponent, potential surface runoff coefficient, snowmelt threshold temperature and degree-day coefficient, frozen soil moisture and temperature), soil erosion and sediment transport parameters (soil erodibility and practice factor in the universal soil loss equation, stream flow peak rate adjustment factor, critical velocity for channel erosion, constant and exponent for calculating the maximum amount of sediment that can be transported in a reach segment), and nutrient yield parameters (initial soil NO_3 and soluble P concentration, organic N and organic P enrichment ratio, phosphorous soil partitioning and percolation coefficient, and nitrate percolation coefficient). Other parameters were set to their default values and were not adjusted during the process of model calibration. Because the purpose of this paper is to introduce the IMWEB-Wetland model development and its modelling ability, detailed descriptions of model calibration for the study watershed are not given here.

For calibration of flow, sediment, and nutrient parameters, model performance was evaluated graphically and statistically based on model bias (BIAS), Nash–Sutcliffe coefficient (NSC), root mean square error (RMSE), and correlation coefficient (CORR) at the EC9 station. BIAS can be expressed as the relative mean difference between observed and predicted results reflecting the ability of reproducing water, sediment and nutrient balance. NSC describes how well the predictions are produced by the model that is commonly used for model evaluation [33]. The calibration objective for flow was to maximize NSC and CORR while simultaneously attempting to reduce BIAS. Calibration of sediment, total nitrogen (TN) and total phosphorous (TP) were conducted for their loadings on sampling days. Observed sediment loading was calculated by multiplying observed sediment concentration by observed flow of the day. Observed TN and TP loadings were calculated by multiplying sampled TN and TP concentrations by observed flow of the day. The model was calibrated firstly for stream flow, then sediment, and finally TN and TP. A summary of model performance at EC9 station for the 2009–2013 calibration period is provided in Table 4, and a graphical comparison between observed and simulated flow at the EC9 station is shown in Figure 7. Overall, the IMWEBs-Wetland simulated stream flows, sediment and nutrient loadings at the outlet station matched the observed data reasonably well based on the statistical assessment results and graphical comparisons.

Under existing condition over the period of 2009–2013, the model predicted an average runoff of 127 mm/year with a runoff coefficient of 0.24, sediment loading 0.04 t/ha, TN loading 2.56 kg/ha, and TP loading 0.47 kg/ha at the watershed outlet (Table 5). Figure 8 shows the simulated spatial distribution of TP yield in the study watershed for the year 2010. Clearly, higher TP losses were from crop lands over the watershed, in which the highest TP losses were in areas with steep slopes. A spatial distribution of wetland TP concentration for the year 2010 is given in Figure 9. The simulated TP concentration was highly variable among the existing wetlands ranging from 0.0 to 0.15 mg/L depending upon their geometric characteristics, the size of upstream contribution areas and their land management practices. However, because no wetland monitoring data were available in the study watershed, the model calibration was conducted at the watershed outlet but not at specific wetland sites. This would cause uncertainties for the wetland modelling results due to input data, model structure, and model parameter estimation.

Table 4. Model performance at EC9 station for the 2009-2013 calibration period.

Item	Samples	Mean	BIAS	NSE	RMSE	CORR
Flow (m^3/s)	2009–2013	0.06	0.01	0.69	0.11	0.83
Sediment loading (t/day)	9	0.09	0.03	0.71	0.05	0.85
TN loading (kg/day)	59	59.9	−0.05	0.56	62.2	0.75
TP loading (kg/day)	59	17.8	−0.03	0.58	20.1	0.85

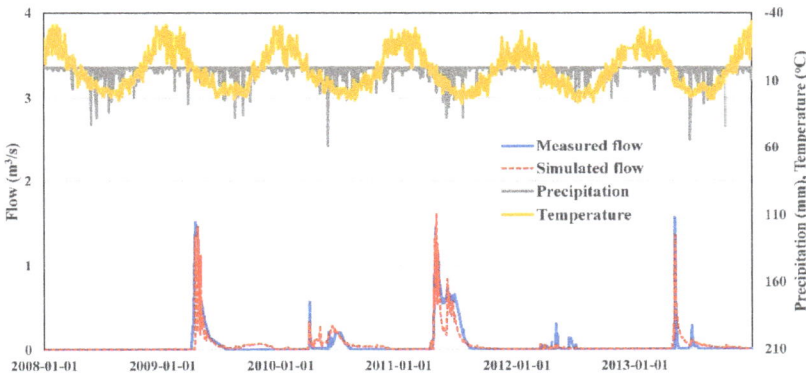

Figure 7. Observed and simulated flow at the EC9 monitoring station.

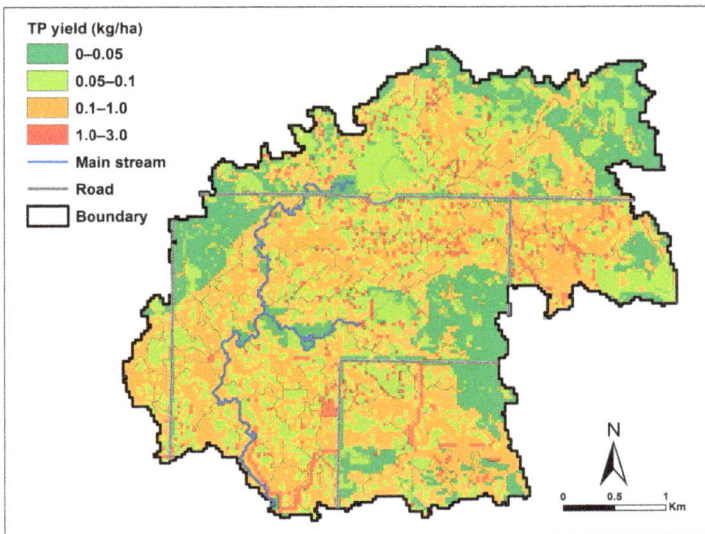

Figure 8. Simulated spatial distribution of TP yield in the study watershed for the year 2010.

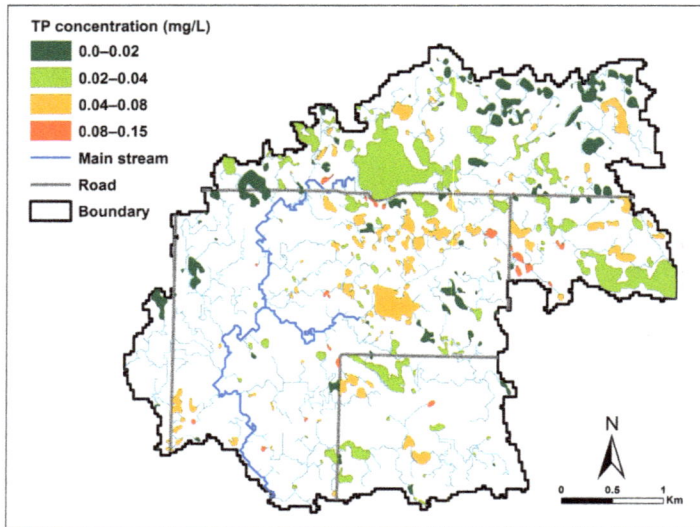

Figure 9. Simulated spatial distribution of wetland TP concentration for the year 2010.

4.3. Scenario Development and Assessment

To demonstrate the ability of the IMWEBs-Wetland model for assessing the effects of wetlands on runoff, sediment and nutrient yields, five scenarios were constructed as listed in Table 5. Scenario I is the baseline scenario with all existing wetlands and existing land management practices. Scenario II is an extreme scenario assuming all existing wetlands are drained and lost for cultivation. Scenario III is another extreme scenario assuming all drained–lost wetlands are restored. Scenario IV and V are two spatial targeting scenarios for TP reduction to identify the top 10 most effective wetlands from the existing wetlands for retention and to identify the top 10 most effective ones from the drained–lost wetlands for restoration in the study watershed. The selection of these wetlands was performed by calculating the TP reduction efficiency with the equation:

$$TP\_E = TP\_R/Wet\_A, \tag{6}$$

where TP_E is the TP reduction efficiency of the wetland (kg/year/ha), TP_R is the annual average wetland TP reduction (kg/year) calculated by subtracting outflow TP from inflow TP of the wetland modelling outputs, and Wet_A is the wetland surface area (ha). The selected 10 most effective existing wetlands for retention and 10 most effective drained–lost wetlands for restoration are shown in Figure 10.

Table 5. Wetland loss and restoration scenarios for the study watershed.

Scenario	Wetland Number	Surface Area (ha)	Storage (10^4 m^3)	Runoff (mm/year)	Sediment (t/year)	TN (kg/year)	TP (kg/year)
I	275	320	123	127	61.0	4020	743
II	0	0	0	154	82.6	4950	896
III	474	500	182	103	42.8	3310	635
IV	265	300	114	132	64.7	4180	767
V	285	371	145	117	53.1	3690	696

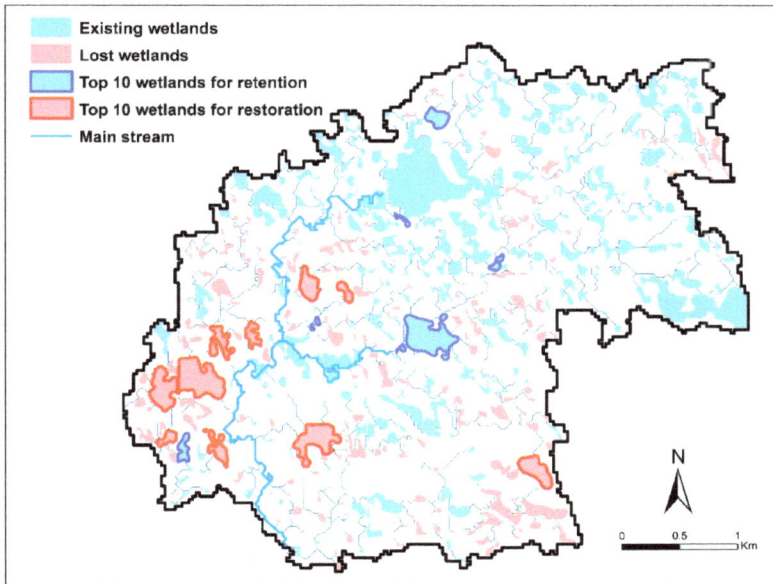

Figure 10. Simulated top 10 wetlands for retention and top 10 wetlands for restoration.

The assessment of wetland loss/restoration impacts on runoff, sediment and nutrient loading for the study watershed was performed based on the 2009–2013 climate and land management data and by changing the wetland operation year in the wetland BMP database. Modelling results were compared to the baseline scenario and are provided in Tables 5 and 6, respectively. Scenario II (loss of all existing wetlands) would increase total runoff at the watershed outlet by 21.3%, sediment by 35.4%, TN by 23.1%, and TP by 20.6% respectively. Scenario III (restoration of all drained–lost wetlands) would decrease total runoff at the watershed outlet by 18.9%, sediment by 29.8%, TN by 17.7%, and TP by 14.5%, respectively.

Table 6. Runoff, sediment, TN and TP changes at the watershed outlet for different scenarios.

Scenario	Runoff		Sediment		TN		TP	
	(mm/year)	(%)	(t/year)	(%)	(kg/year)	(%)	(kg/year)	(%)
II	27.0	21.3	21.6	35.4	930	23.1	153	20.6
III	−24.0	−18.9	−18.2	−29.8	−710	−17.7	−108	−14.5
IV	5.00	3.94	3.70	6.07	160	3.98	24.0	3.23
V	−10.0	−7.87	−7.90	−13.0	−330	−8.21	−47.0	−6.33

Scenario IV (loss of 10 most effective wetlands) would increase runoff by 3.94%, sediment by 6.07%, TN by 3.98%, and TP by 3.23%, respectively. Scenario V (restoration of 10 most effective wetlands) would decrease runoff by 7.87%, sediment by 13.0%, TN by 8.21%, and TP by 6.33% respectively. The full wetland loss scenario (II) and full restoration scenario (III) have average TN and TP increase rates of 2.91 and 0.48 kg/ha and average TN and TP reduction rates of 3.94 and 0.60 kg/ha, respectively. The relatively small reduction rates of TN and TP after full restoration (Scenario III) are associated with different N and P forms, for which part of dissolved N and P percolated from the wetlands would return to mainstreams with groundwater flow. As a result, the reduction rates for runoff, TN, and TP are on the same order and smaller than the sediment reduction rate based on model simulation

(Table 6). In contrast, Scenario IV (loss of top 10 most effective wetlands for retention) and Scenario V (restoration of the 10 most effective wetlands) have average TN and TP increase rates of 8.0 and 1.2 kg/ha and average TN and TP reduction rates of 6.47 and 0.92 kg/ha, respectively. Considering the hundreds of existing and drained–lost wetlands in the study watershed, spatial targeting of those wetlands with higher TP reduction rates for retention and restoration has the potential of improving the effectiveness of wetland conservation programs.

5. Discussion and Conclusions

A cell-based modular modelling system, IMWEBs-Wetland, is developed for simulating and assessing the water quantity and water quality effects of individual wetlands at a watershed scale. The model is supported by one modular library and five databases (geospatial, hydro-climate, BMP, parameter, and output), which are managed by a Whitebox GAT based user interface. The model simulates processes of climate, flow, sediment, and water quality by incorporating land management practices. Compared to other watershed wetland models, the IMWEBs-Wetland model has distinguishing features of: (1) setting up the model based on project objective, watershed characteristics, data availability, and outputs with interest; (2) simulating hydrologic processes with more spatial details and providing both time series and spatial distribution outputs at user-defined spatial and temporal scales; (3) integrating with economic and ecologic models more easily for cost-effective assessment of wetland loss and restoration scenarios due to its cell based structure; and (4) interfacing with an open-source Whitebox GIS and SQLite database. However, some limitations also exist for the IMWEBs-Wetland model. For example, the model needs more detailed BMP distribution and operation data for model setup, and site-specific observation data for model calibration. Because the model runs for each grid cell, it would take a long computational time and a large memory space for a watershed with a small cell size.

A case study of IMWEBs-Wetland modelling was conducted in a small watershed in southern Manitoba of Canada with hundreds of existing and drained–lost wetlands. The model was setup based on existing wetlands and land management conditions. Calibration results demonstrated that the model performed well for flow, sediment, and water quality simulation at the watershed outlet. A simulated TP spatial distribution showed that TP concentrations were highly variable among existing wetlands in the study watershed depending wetland and its contribution area characteristics. However, because the model was not calibrated at wetland sites, this may cause uncertainties in the modelling results. Four wetland loss and restoration scenarios were constructed and evaluated with the calibrated IMWEBs-Wetland model. Compared to Scenario II (loss of all existing wetlands) and Scenario III (restoration of all drained–lost wetlands), the spatial targeting scenario IV (retention of the top 10 existing wetlands with the highest TP reduction efficiency) and scenario V (restoration of the top 10 drained–lost wetlands with the highest TP reduction efficiency) are more effective in reducing sediment, TN, and TP yields at the watershed outlet for wetland retention and restoration. As indicated by Fossey et al. [13] and Evenson et al. [15], field monitoring data are critical in validating distributed models and applying modelling results for watershed management. Because no monitoring data were available to validate the model at representative wetland sites in this study, aggregated calibration was conducted at the watershed outlet and uncertainties of modelling results would exist for individual wetlands. Therefore, site-specific observation data for enhanced model calibration are essential for improving the reliability of modelling results.

With a cell-based structure, modular system, open-source GIS and database, and the use of advanced modelling techniques, the IMWEBs-Wetland model is capable of modelling and assessing wetland and other BMP scenarios for spatial watershed management and decision-making. The modelling results have the potential to improve the effectiveness of wetland conservation programs. Future research and development perspectives of the model include: (1) improvement of the model by developing new algorithms for specific hydrologic process modules; (2) improvement of the model by adding more BMP modules; (3) improvement of the model to perform an effective sensitivity analysis, auto-calibration, uncertainty assessment, and spatial optimization; and (4) integration with economic and ecologic models for cost-effective analysis and assessment of wetland and other BMPs in a watershed; and (5) validation of the model in watersheds with different climate and landscape conditions.

Author Contributions: Y.L., W.Y., H.S. and Z.Y. conducted the model design and development. Y.L., W.Y., H.S. conducted the case study and paper writing. J.L. significantly contributed towards writing and editing the manuscript and also provided valuable guidance throughout the study.

Funding: This research was funded by Agriculture and Agri-Food Canada under agreement No. 1585-10-3-2-1-2, Ontario Ministry of the Environment and Climate Change under agreement No. 1314017, Alberta Agriculture and Forestry, Alberta Innovates Bio Solutions, and InnoTech Alberta under agreement No. G2015C00303, and Alberta Biodiversity Monitoring Institute under agreement No. ABMI-14-39.

Acknowledgments: The IMWEBs model development originated from the Watershed Evaluation of BMPs (WEBs) program in Agriculture and Agri-Food Canada (AAFC) and was further supported by Ontario Ministry of the Environment and Climate Change (MOECC), Alberta Agriculture and Forestry, Alberta Innovates Bio Solutions, and InnoTech Alberta. We would like to thank Terrie Hoppe and Brook Harker of AAFC, Jinliang Liu of MOECC, Clinton Dobson of Alberta Agriculture and Forestry, Carol Bettac of Alberta Innovates Bio Solutions, and Marian Weber of InnoTech Alberta for their project management support on IMWEBs development. The development of IMWEBs wetland module was supported by Alberta Biodiversity Monitoring Institute (ABMI). We would like to thank Dan Farr, Eric Butterworth, and Tom Habib of ABMI for their project management support on IMWEBs-Wetland development. We would also like to thank Pascal Badiou, Lyle Boychuck, Bryan Page, and Shane Gabor of Ducks Unlimited Canada for providing valuable datasets of the study site.

Conflicts of Interest: The authors declare no conflict of interest.

References

1. Golden, H.E.; Lane, C.R.; Amatya, D.M.; Bandilla, K.W.; Kiperwas, H.R.; Knightes, C.D.; Ssegane, H. Hydrologic connectivity between geographically isolated wetlands and surface water systems: A review of select modeling methods. *Environ. Model. Softw.* **2014**, *53*, 190–206. [CrossRef]

2. Wang, X.; Yang, W.; Melesse, A.M. Using hydrologic equivalent wetland concept within SWAT to estimate streamflow in watersheds with numerous wetlands. *Trans. ASABE* **2008**, *51*, 55–72. [CrossRef]

3. Quinton, W.L.; Hayashi, M.; Pietroniro, A. Connectivity and storage functions of channel fens and flat bogs in northern basins. *Hydrol. Process.* **2003**, *17*, 3665–3684. [CrossRef]

4. Yang, W.; Wang, X.; Liu, Y.; Gabor, S.; Boychuk, L.; Badiou, P. Simulated environmental effects of wetland restoration scenarios in a typical Canadian prairie watershed. *Wetl. Ecol. Manag.* **2010**, *18*, 269–279. [CrossRef]

5. Yang, W.; Liu, Y.B.; Cutlac, M.; Boxall, P.; Weber, M.; Bonnycastle, A.; Gabor, S. Integrated economic-hydrologic modeling for examining cost-effectiveness of wetland restoration scenarios in a Canadian prairie watershed. *Wetlands* **2016**, *36*, 577–589. [CrossRef]

6. Mekonnen, B.A.; Mazurek, K.A.; Putz, G. Incorporating landscape depression heterogeneity into the Soil and Water Assessment Tool (SWAT) using a probability distribution. *Hydrol. Process.* **2016**, *30*, 2373–2389. [CrossRef]

7. Mekonnen, B.A.; Mazurek, K.A.; Putz, G. Sediment export modeling in cold-climate prairie watersheds. *J. Hydrol. Eng.* **2016**, *21*, 05016005. [CrossRef]

8. Nasab, M.T.; Singh, V.; Chu, X. SWAT modeling for depression-dominated areas: How do depressions manipulate hydrologic modeling? *Water* **2017**, *9*, 58. [CrossRef]

9. Martinez-Martinez, E.; Nejadhashemi, A.P.; Woznicki, S.A.; Adhikari, U.; Giri, S. Assessing the significance of wetland restoration scenarios on sediment mitigation plan. *Ecol. Eng.* **2015**, *77*, 103–113. [CrossRef]

10. Abouali, M.; Nejadhashemi, A.P.; Daneshvar, F.; Adhikari, U.; Herman, M.R.; Calappi, T.J.; Rohn, B.G. Evaluation of wetland implementation strategies on phosphorus reduction at a watershed scale. *J. Hydrol.* **2017**, *552*, 105–120. [CrossRef]

11. Hattermann, F.F.; Krysanovaa, V.; Habecka, A.; Bronstert, A. Integrating wetlands and riparian zones in river basin modelling. *Ecol. Model.* **2006**, *199*, 379–392. [CrossRef]

12. Hattermann, F.F.; Krysanova, V.; Hesse, C. Modelling wetland processes in regional applications. *Hydrol. Sci. J.* **2008**, *53*, 1001–1012. [CrossRef]

13. Fossey, M.; Rousseau, A.N.; Savary, S. Assessment of the impact of spatio-temporal attributes of wetlands on stream flows using a hydrological modelling framework: A theoretical case study of a watershed under temperate climatic conditions. *Hydrol. Process.* **2015**, *30*, 1768–1781. [CrossRef]

14. Fossey, M.; Rousseau, A.N.; Bensalma, F.; Savary, S.; Royer, A. Integrating isolated and riparian wetland modules in the PHYSITEL/HYDROTEL modelling platform: Model performance and diagnosis. *Hydrol. Process.* **2015**, *29*, 4683–4702. [CrossRef]

15. Evenson, G.R.; Golden, H.E.; Lane, C.R.; D'Amico, E. Geographically isolated wetlands and watershed hydrology: A modified model analysis. *J. Hydrol.* **2015**, *529*, 240–256. [CrossRef]

16. Evenson, G.R.; Golden, H.E.; Lane, C.R.; D'amico, E. An improved representation of geographically isolated wetlands in a watershed-scale hydrologic model. *Hydrol. Process.* **2016**, *30*, 4168–4184. [CrossRef]

17. Golden, H.E.; Sander, H.A.; Lane, C.R.; Zhao, C.; Price, K.; D'amico, E.; Christensen, J.R. Relative effects of geographically isolated wetlands on streamflow: A watershed-scale analysis. *Ecohydrology* **2016**, *9*, 21–38. [CrossRef]

18. Stuart, V.; Harker, D.B.; CLearwater, R.L. *Watershed Evaluation of Beneficial Management Practices (WEBs): Towards Enhanced Agricultural Landscape Planning-Four-Year Review*; Agriculture and Agri-Food Canada: Ottawa, ON, Canada, 2010.

19. Liu, Y.; Yu, Z.; Chung, S.; Lung, I.; Yang, W.; Yarotski, J.; Elliott, J. *Developing a Modular, Fully Distributed Hydrologic Model for Examining Place-Based BMP Effects in the Steppler and South Tobacco Creek Watersheds*; University of Guelph: Guelph, ON, Canada, 2013.

20. Liu, Y. Development of a GIS-Based Hydrologic Model for Flood Prediction and Watershed Management. Ph.D. Dissertation, Free University of Brussels, Brussels, Belgium, 2004.

21. Arnold, J.G.; Moriasi, D.N.; Gassman, P.W.; Abbaspour, K.C.; White, M.J.; Srinivasan, R.; Santhi, C.; Harmel, R.D.; Van Griensven, A. SWAT: Model use, calibration, and validation. *Trans. ASABE* **2012**, *55*, 1491–1508. [CrossRef]

22. Wang, X.; Williams, J.R.; Gassman, P.W.; Baffaut, C.; Izaurralde, R.C.; Jeong, J.; Kiniry, J.R. EPIC and APEX: Model use, calibration, and validation. *Trans. ASABE* **2012**, *55*, 1447–1462. [CrossRef]

23. Lowrance, R.; Altier, L.S.; Williams, R.G.; Inamdar, S.P.; Sheridan, J.M.; Bosch, D.D.; Hubbard, R.K.; Thomas, D.L. REMM: The riparian ecosystem management model. *J. Soil Water Conserv.* **2000**, *55*, 27–34.

24. Wiens, L.H. A surface area-volume relationship for Prairie wetlands in the Upper Assiniboine River Basin, Saskatchewan. *Can. Water Resour. J.* **2001**, *26*, 503–513. [CrossRef]

25. Neitsch, S.L.; Williams, J.R.; Arnold, J.G.; Kiniry, J.R. *Soil and Water Assessment Tool Theoretical Documentation Version 2009*; TR-406; Texas A&M University System: College Station, TX, USA, 2011.

26. Lindsay, J.B. Whitebox GAT: A case study in geomorphometric analysis. *Comput. Geosci.* **2016**, *95*, 75–84. [CrossRef]

27. Soille, P.; Gratin, C. An efficient algorithm for drainage network extraction on DEMs. *J. Vis. Commun. Image Represent.* **1994**, *5*, 181–189. [CrossRef]

28. O'Callaghan, J.; Mark, D. The extraction of drainage networks from digital elevation data. *Comput. Vis. Graph. Image Process.* **1984**, *28*, 323–344. [CrossRef]

29. Manitoba Agriculture. State of Agriculture in Manitoba. Available online: http://www.gov.mb.ca/agriculture/markets-and-statistics/yearbook-and-state-of-agriculture/pubs/state_of_ag_pubn.pdf (accessed on 12 June 2018).

30. Honey, J. *Crops in Manitoba 2012*; Department of Agribusiness and Agricultural Economics, University of Manitoba: Winnipeg, MB, Canada, 2013.

31. NASA. NASA Prediction of Worldwide Energy Resource (POWER) POWER Data Access Viewer. Available online: https://power.larc.nasa.gov/data-access-viewer/ (accessed on 12 June 2018).

32. Energy, Mines, and Resources Canada. Solar Radiation—December and June. Available online: http://www.nlcpr.com/images/CanadaSolarDecJune_2M.jpg (accessed on 12 June 2018).

33. Nash, J.E.; Sutcliffe, J.V. River flow forecasting through conceptual models, Part I—A discussion of principles. *J. Hydrol.* **1970**, *10*, 282–290. [CrossRef]

MDPI

St. Alban-Anlage 66

4052 Basel

Switzerland

Tel. +41 61 683 77 34

Fax +41 61 302 89 18

www.mdpi.com

Water Editorial Office

E-mail: water@mdpi.com

www.mdpi.com/journal/water

www.ingramcontent.com/pod-product-compliance
Lightning Source LLC
Chambersburg PA
CBHW051907210326
41597CB00033B/6061